# Bear Trap Mountain

Daniel Hance Page

First printing



ISBN: 978-1-61296-435-5

PUBLISHED BY BLACK ROSE WRITING

www.blackrosewriting.com

Printed in the United States of America

Suggested retail price $15.95

*Bear Trap Mountain* is printed in Palatino Linotype

FOR

Marg, Hank, Jim, Sheldon, Shane and Shannon Page, John and Dan Robinson, Lester and Rose Anderson, Joe and Linda Hill, Macari Bishara, Kevin and Alison Griffin, Jerry and Gaye McFarland, "Mac" McCormick, Grant Saunders, Frank Lewis and other friends with whom we have enjoyed both the wilderness in the forest by Parry Sound, Shebeshekong River, Bremner River and Quinlan's woods and also the wilderness where the ocean touches the shore at Belleair Beach.

# Bear Trap Mountain

*"If a man walk in the woods for the love of them half of each day, he is in danger of being regarded as a loafer; but if he spends his whole day as a speculator, shearing off those woods and making earth bald before her time, he is esteemed an industrious and enterprising citizen. As if a town had no interest in its forests but to cut them down."*

**~Henry David Thoreau, 1854**

# PART ONE

The mayor and councilors of Clarksville did not realize their town was about to be invaded when the old man stood up to speak to the council. If the man was trying to be inconspicuous, the ruse worked because no one in the council particularly acknowledged the stranger's presence. On the agenda, he was listed as Caleb Pine who lived on Bear Trap Mountain with his dog that was a coyote. The visitor's hair was white and reached the shoulders of his black jacket topping a plaid shirt and vest. He also wore jeans and boots. A strand of white hair dropped across his broad forehead. His nose was finely chiseled above a chin and jaw that were both strong. Grayish blue eyes sparkled behind silver rimmed glasses.

The mayor sat in his usual place, as did each councilor. Their disinterest put a lifeless, gray tone to a bureaucratic routine that made no concession to excitement. Some members drank coffee although even this stimulating brew did not stir any apparent interest in the present speaker.

Speaking in a low volume, his voice had a deeply intoned ring that had some affect on these people habitually cloaked behind a gray film of disinterest. The man said, "At the east end of town, there is a large area of rubble bordered by the Sand River to the south, River Street on the west, Church Street along the north and Arrowhead Road to the east. I would like to purchase this land from the town."

"Why would you want to buy such a wasteland?" asked a councilor.

"I would like to grow and sell farm produce as well as make and sell works of art," answered the man. The mayor and councilors laughed and their agreement to sell the land was unanimous.

Leaving the Council Chamber, Caleb walked to some stairs and hurried down them. Air was stale and permeated with an odor of decaying wood. Proceeding to large, front doors, he opened one and stepped outside the building, welcoming a cool, early spring morning. A robin sang from the branch of a maple. Another robin pulled a worm from a thin layer of snow covering grass.

A bird's song makes life beautiful even in this old town, thought Caleb while he noted the derelict appearance of buildings where businesses had thrived in previous years. Most industries had moved away, leaving behind only the pollution. A chemical odor tainted a breeze blowing from the west.

A few days later, after transactions for the land purchase had been completed, Caleb walked westward until he came to the ramshackle forms of factories. Smoke stacks stood starkly against a bluish sky. Gray smoke spewed from these stacks and formed an horizontal, westerly drifting cloud.

Fences bordered and secured the industrial zone. Caleb walked beside a fence and approached the bank of the Sand River. Clean water had previously flowed here, noted Caleb. This river is now just a stream of garbage and a fire hazard. Oily substances swirl where once there was pure water.

Arriving at the entrance to a large, drainage pipe, he stepped inside. This pipe drains water from the road, he thought. The road runs beside the fence.

He proceeded onward inside the tunnel. Splashing sounds, from his boots, echoed within the narrow channel, disturbing numerous rats. They scrambled ahead, churning the watery floor. Streaks of light pierced the murkiness from metal covers providing access to the road.

The underground passageway led to a spacious, rectangular chamber formed by axe-hewn timbers. The floor of this chamber was covered by sand and gravel topped by an oily liquid. Large, gaping pipes entered from the sides. Caleb stepped into one of the pipes that led westward.

He walked until he knew he was inside the fence and near the buildings. He pushed on a metal grate at the top of the pipe. The grate moved upward allowing sunlight to enter the tunnel.

Peering out of the opening, Caleb saw that he was in the center of a service road close to a factory. When a truck loomed into view, he dropped the grate just before rumbling sounds followed the truck passing overhead along the road.

Returning back through the tunnel, Caleb stepped out of the pipe amid golden hues from the setting sun. He enjoyed the panorama of the sun's deepening color while descending into pink clouds at the horizon. Clouds became emblazoned in scarlet before this light faded. Afterward, evening shadows crept slowly across the land.

Satisfied that conditions were favorable, Caleb switched on his flashlight then entered the pipe and started retracing his route toward the factories. Rats scampered ahead of the moving beam of light. Outside the grate, near the factories, clouds drifted westerly. The flight of an hunting, great horned owl left no shadow upon the land.

Seeing the metal cover move, the owl circled back and flew to the top of a post. From this location, the grate could be watched along with the road and buildings. The grate fell backwards clunking onto the road. In the uncovered opening, the old man appeared. He held a wooden bow in addition to arrows. The arrows, one at a time, became tipped with fire before being pulled back in the bow and shot into the night sky. One bolt of light after another, issuing whooshing sounds, arced toward the buildings. Sounds of breaking glass rattled through the murkiness as flaming arrows entered the factories. Lights flashed inside the structures. After the old man reentered the opening and replaced the grate, the owl left the post and flew into a murky sky that was brightening around a tower of roaring flames.

Caleb stepped out of the pipe and started walking westward along a garbage-strewn bank of the river oozing with oily slime. He progressed slowly until reaching the main tributary. Where spring water had at one time entered the Sand River, there was now a slowly flowing oily substance. Into outlet pipes, Caleb shot flaming arrows, starting a rumbling of underground explosions. The earth shook and fireballs shot from the pipes sending flames whooshing along the river including its tributaries. While a flash of flame followed the river, explosions continued to shake the ground until they joined into one, thundering rumble under a fireball that lit up the sky. As flames traveled, the patterns of fires widened until the polluted areas of Clarksville became an inferno. Thundering and exploding fires obliterated most of the sounds of sirens from police cars and fire trucks.

Caleb hurried westward to check the progress of the fires. Along Westbury, as well as Main Street, the businesses and houses were unharmed. Beyond Main Street, fires raged in the old, polluted, industrial section of town. A drifting haze of smoke was touched by firelight.

Returning eastward, Caleb saw flames shooting from the abandoned factories on the far side of River Street. South of town, flames had jumped from the burning river to ignite grass fires. They spread in a southeasterly direction, consuming brush and abandoned, farm buildings. This traveling fire enveloped shells of buildings on the southern part of the recently purchased land. East of Arrowhead Road, flames removed derelict factories and apartment buildings.

The fire has cleared away the ruined parts of town, noted Caleb with satisfaction while he stood beside a pine on the highest part of his new property. This lofty pine stretched branches into a night sky containing acrid smoke.

Walking northward, on his property, Caleb came to an extensive section of hard maples. This is an healthy stand of trees, he noted before proceeding onward to Church Street then walking westward until he reached Main Street.

Everyone in town is out watching the fires, thought Caleb. Cars clogged the streets. Sidewalks were bustling with people. Flames emblazoned the sky farther to the west. Caleb approached a policeman who was sipping coffee from a paper cup while watching the surrounding action. The man was tall with gray hair and black eyes.

"What has been happening?" asked Caleb.

"The old parts of town have been fire hazards for years,"

answered the man. "These oily places have finally caught on fire. Even the river's burning. No one has been injured. Most stores are open. The fires are restoring some business activities."

"The fire will clean the town," said Caleb.

"Yes," replied the policeman. "Even the river was a fire hazard."

Turning to leave, Caleb said, "If the stores are open, I should get some supplies."

"Good idea," replied the policeman. "This town has come back to life."

Caleb walked to an hardware store with a lumber yard. He ordered land-clearing tools including an axe, pickaxe, crowbar and shovel. He also selected materials for building a lean-to style of cabin beside the tall pine on his property. The purchase of cooking equipment and food supplies completed the night's work. He operated out of a motel until the fires had burned out then carried bags of food to his new campsite. Other items were delivered.

Near the base of the white pine, Caleb kindled a fire. With his back leaning against the trunk, he slept until dawn brought the first, gray light to the hilltop.

Wood added to coals brought flames to the fire pit, adding warmth to dispel the coolness of the morning. Overhead, long flocks of geese flew northward. Their resonant calls spoke of new life to the smoldering landscape. Purple finches, along with a cardinal, sang from the maple grove. These songs were occasionally pierced by the shrill call of a blue jay. Sounds of spring, mused Caleb.

A crow, carrying a stick in her beak, flew to a nest being built near the top of the pine. A second crow arrived with an additional twig. I suppose, in our own ways, we all build nests, reflected Caleb as he worked at his lean-to shelter.

Caleb fried bacon in a pan over coals. Potatoes were boiled before being cut into thick slices and fried briefly. Lastly, he perked coffee in his battered and blackened pot.

After breakfast, while sipping coffee, Caleb observed his

smoldering property. Broken walls of collapsed buildings jutted bleakly against the skyline. Roughly mortared, stone foundations pitted the earth. Much of the area was a blackened wasteland. Fires continued to flicker, adding tendrils of smoke to dawn air. The tallest structure remaining to mark the skyline was a tall building being constructed on the east side of Arrowhead Road and north of the Sand River. This building site had not been touched by the onrush of fires that left behind a charred landscape. Ashes will fertilize new growth, noted Caleb. The earth will send forth all new shoots and be restored.

I forgot to feed the birds, noted Caleb. He walked to a feeder he had placed south of his lean-to. On this feeder, he dropped pieces of bread and fried potatoes along with peanuts. Crows and blue jays flew to the platform while Caleb refilled his coffee cup and sat down again to enjoy warmth from the fire.

Southeast of the lean-to, a section of asphalt had buckled. Caleb selected his new pickaxe and walked to the lowest section of pavement. He swung the axe and it sank into the asphalt. Additional chops broke away a thick layer, exposing a second surface. Methodically, this layer was also removed before Caleb rested.

He returned to the lean-to and sat down with his back leaning against the pine's trunk. Flakes of snow drifted from a mottled sky. When sunlight flashed through breaks in clouds, pools of light moved across the landscape. Sporadic sounds from motors of equipment came from the tall building on the far side of Arrowhead Road. A crow cawed while flying to the top of the white pine. A cool day, like today, noted Caleb, is a fine time to work.

He picked up his axe and shovel then walked to the hole he had chopped through two layers of asphalt. Beneath this pavement, there was gravel. By first using the axe to loosen the gravel, Caleb removed it with the shovel. Digging continued, in a daily routine, until the top and sides of a boulder had been uncovered.

During break periods, he prepared green tea. Sitting down

on top of the boulder, he sipped the drink contentedly while watching the surrounding landscape. As sea gulls soared overhead, sunlight touched their feathers etching them brightly against a backdrop of an azure sky. I'll have to cut a path in order to roll this rock downhill to the large foundation, Caleb said to himself.

By the time the sun had moved across the sky on the last of a few days of work, Caleb Pine had cleared away the buckled pavement along with the gravel. He had also cleared a pathway downhill for the boulder. The land was being painted by red hues from the setting sun when he decided to stop working and return to his lean-to.

A robin sang while Caleb prepared a meal of fried bread for himself and the crows. After enjoying a cup of tea, he added wood to the fire then slept in the lean-to under a starlit sky.

When first rays of sunlight flashed across the land, songs of robins filled the morning. These melodies were occasionally joined by harsher calls of grackles and crows.

Caleb wedged a crowbar under the boulder and pushed on the bar with unusual force. The rock moved, issuing a grating sound. The bar was repositioned and used again to dislodge the boulder. A third thrust with the lever pushed the giant from its resting place. Seeming to hesitate with indecision, an additional push moved the rock forward until it toppled onto the southeastern trench. While rolling, speed increased then the boulder bounced and hit the foundation's edge before jumping upward, entering the air with an eerie silence while arcing into the cavity. After a sharp crashing sound, silence returned.

The rock's removal from the earth left a rounded hole that started to fill with water. While Caleb watched, clear water flowed upward and gradually filled the space where the rock had been. I dislodged a boulder and it became an almost unstoppable force, noted Caleb. The stream has now become unblocked and it will be unstoppable.

Clear water trickled from the edge of a pool and proceeded along the path taken by the boulder. This freshly released rivulet reached the foundation then splashed down to its floor.

After hitting the floor, the boulder had rolled to the eastern wall. At the point of impact, a large hole had been cut through the cement. The stream of water entered this opening.

Caleb used his axe to cut notches into an old beam in order to make a ladder. He angled the beam from the top of the wall to the foundation's floor. Carrying an assortment of tools, he stepped down along the ladder.

A particularly large crack had spread out from the main break caused by the boulder's impact. The stream of water, entering the central hole, dropped into a lower cavity. Caleb first used his pickaxe to widen the break in the cement. Beneath this floor, he uncovered a second, older layer of cement. This surface was also cracked. The rivulet of spring water entered a fissure in the old cement and fell into a lower chamber.

The pickaxe was used again to enlarge the fissure. Chunks of cement fell, with an echoing crash, into a cavern. Caleb worked through his daily routine until he had chopped an opening he could enter.

After returning to his lean-to to get his longest rope, Caleb tied it to a beam protruding from a wall. He dropped the rest of the rope into the opening in the floor. With a pickaxe in his belt, along with a flashlight, he held the rope and stepped into the hole. He descended into a murky chamber where some light entered through the break. He descended until his boots touched a wet, cement floor of an old tunnel. Walls were constructed of stones held in place by coarse mortar. Air was stale, damp and cold. A stream of water fell from the overhead opening. The water covered a cement floor and flowed northward over a stone passageway. Something moved on the wall, sending a pebble splashing into water covering the floor. The flashlight's beam of light revealed part of a snake slithering along a break in the wall. Other snakes rustled between gaps in mortar or moved across the watery floor.

Must be an hibernation area, Caleb said to himself. Snakes are starting to stir about. An ominous rattling sound echoed in the murky confines of the tunnel while water continued to flow over the floor. Caleb followed the rippled stream.

A gurgling sound marked the location of three stone steps. The creek trickled down them then entered the smooth surface of a water-covered floor. A slight current rippled the surface and turned toward the center of the floor beneath the northern end of a cavern. In the western corner, of this chamber, there was a vault sided by stones held in place by the coarse mortar. The vault's eastern wall extended from the floor to the dome-shaped ceiling. Much of the front of the structure consisted of a rusted door. A stone shelf protruded across the back of the cavern from the vault's side to the eastern wall. On this ledge, there were rusted metal boxes.

Diverging ripples moved away from a snake's head as it zigzagged across water covering the floor. Pieces of mortar dropped into the water when snakes slithered in the walls. After watching the snake's head cross the floor, Caleb noticed that the water level was rising. I unblocked the spring, noted Caleb and its water is going to fill this tunnel. I'm running out of time.

He walked to the rust-encrusted door and was unable to move its handle. Failing to dislodge the door, he looked back at the hole he had cut through the ceiling. Dim light entered through this break, as did a stream of water. Splashing sounds echoed in the tunnel, reminding Caleb of the passing of valuable time. Air continued to be humid, stale and cold. Snakes moved although most of them seemed to have retreated into holes in the walls.

The weakest part of the vault will be its eastern wall, reasoned Caleb. Stones are held in place by mortar that will have deteriorated through the years. My main concern is the rising water level in this tunnel, he said to himself. It is filling gradually. I could try to block the spring again or attempt to change the direction of the stream. However, if I work quickly, I'll be able to stay ahead of the water. The snakes are another problem. The water that traps me will also trap them. I'd hate to get trapped in here with snakes. The air is also stale. The amount of water flowing into this chamber might suddenly increase.

Using his flashlight, he more closely checked the vault's wall. The mortar was porous and crumbling. One large rock was bordered by particularly large cracks. He swung his pickaxe and imbedded its point in cement directly above the rock. Chunks of mortar were dislodged and splashed to the watery floor. Additional chunks fell from the ceiling just before something slithery and ropelike dropped on Caleb's neck and shoulders. Screaming in revulsion, he swung his arms and swung a large, writhing snake into the water. More mortar fell from the ceiling as a snake slithered into an upper crevasse between two rocks.

Another chop with the axe sent its tip deeply into crumbling mortar. Chips fell from around the rock and also the ceiling. Chopping continued while snakes moved and the water level stirred menacingly upward.

Having removed much mortar, Caleb wedged the axe's tip into a larger opening and dislodged the rock, sending it splashing into deepening water. Surrounding stones were easier to remove. They revealed a second wall. Continuous chopping loosened more mortar and stone in the new barrier.

The rising water level was approaching the top of Caleb's boots by the time his axe pierced the second wall. He widened this opening until he thought he could crawl through it. Before entering, he used the flashlight to check the vault's interior. Nothing moved inside. At the back of this chamber, there were two stone shelves.

Caleb climbed through the opening. Continuing to check the area with his flashing, he stepped forward to the stone shelf. On top of it, there was a rusted box containing gold coins. Beside this container, there was jewelry in a rotted bag. A lower shelf contained gold bars.

Cal removed his jacket and used it as a carrying bag. He first filled it with jewelry and gold then pushed it through the hole in the wall. After dropping the bag into water, he climbed through the break. Moving slowly because of the bundle's weight, he walked to the steps. The first step was under water.

Pulling the bundle up the steps, he reached the opening in the ceiling where an increased flow of water entered and splashed onto the floor. He tied his pack to the rope before climbing it and scrambling back into daylight. The heavy bundle was slowly pulled up out of the tunnel. He climbed the roughly cut ladder and again used the rope to remove his cargo and take it to the lean-to. He buried the jewelry and gold under the lean-to. Having completed this work, he rested in the shelter then slept until he welcomed a new day with the grayness of dawn.

A light coating of snow outlined the land and gave a sharper edge to the cold. He added pieces of boards to the fire until tall flames flickered into still air. The warmth was welcomed, as was the light.

Caleb carried his coffee pot to the spring. Ice bordered a clear pond. Near its center, a current swirled toward a channel where an increasingly large flow of water splashed into the foundation. Under this clear stream, there gradually appeared sand and smooth pebbles of an old riverbed. Smooth stones also ringed the pond. Cal dipped his pot into the clear, cold water. He filled the pot by moving it in the direction the current was stirring. Back at camp, coffee was soon perking fragrantly. To a blackened frying pan, he added sunflower seed oil then added dough to fry bannock. After one side turned to a golden color, he turned the bread and cooked it again while enjoying the wheaten fragrance.

Breakfast consisted of fried bread topped by honey and coffee. Pieces of bannock were placed on the feeder and the crows came almost immediately to get this food. They drank at the spring, as did many other birds. Every day, more birds arrived to get water at the spring.

Caleb sat down, leaned his back against the pine's trunk then sipped coffee slowly, savoring its rich earthy taste. The drink helped relieve his weariness as well as the cold.

I don't have time to rest, he warned himself. I must return for the remaining bars before water seals the tunnel. Hastening away from the comforts of his camp, he carried his rope, coat and flashlight to the foundation. Having proceeded carefully down the roughly chopped steps in the ladder, he used the rope again to enter the tunnel. Hissing sounds of falling and splashing water filled the chamber. Water was rising much faster than he had expected.

Deep water lurked above his knees. Snakes stirred more than they had previously. Heads moved among ripples. Steps were below the surface although the flashlight's beam of light found them.

He approached the vault and climbed through the hole in the wall. A large snake had taken refuge on the top shelf. The reptile's mouth opened as the thick body writhed into a coil. The head moved upwards then back. Caleb threw his coat, blocked the serpent when it sprang from the shelf then used his flashlight to club his moving coat. He pulled it to the ledge while the snake became dislodged. Part of the long body floated. Most of the creature remained submerged.

Wary of the rising water, Caleb worked quickly to place the remaining bars on his jacket and use it again like a carrying bag. A rat backed into murkiness when Caleb entered the hole and stepped into the tunnel. A snake's head, moving along the water's surface, swerved directly toward Caleb. He swung his flashlight to strike the creature. It thrashed in the water behind Caleb while he moved up the steps. He returned to the rope, climbed it then pulled up his jacket filled with gold bars.

Back at his camp, he buried the rest of the gold under the lean-to. Feeling exhausted, he slept until the next dawn that was again cold. Flakes of snow drifted downward from a gray sky.

For breakfast, he prepared the usual bannock and coffee. While resting and sipping coffee, he decided to walk to town.

Snowflakes drifted with a west wind. Robins and grackles searched for worms while a skiff of snow added distinct outlines to the landscape. I've always appreciated the birds because they eat insects, thought Caleb. I've also been grateful for the insects that feed the birds. In attempts to kill insects, people are destroying the birds' food supply. People are putting too much poison into the earth, water and air. I don't understand why anyone would spray poison onto grass in order to kill insects that birds need for food. Poison kills birds directly while killing people indirectly. If there were more insects, I would hear a whippoorwill sing here again.

Sunlight slanting across the earth gradually melted the dusting of snow. In this warming light, purple finches and robins blended their songs into a cheerful melody. Caleb rested beside his warming fire until the sun had climbed the sky and become obscured by eastward drifting clouds.

He placed the gold and jewelry in an ash, splint basket that he carried on his back. He walked westward to River Street then past the burned remnants of factories. At Westbury Street, there were houses on the west side that had not been touched by fire. Continuing westward, he followed a path to Main Street. He proceeded directly to the bank and entered the building.

Bank employees looked at each other questioningly when they saw the old man walk to the manager's office. The manager, wearing a dark blue suit, had grayish-brown, closely cropped hair. Dark rimmed glasses matched the suit. Working at a desk, he looked up when Caleb's form filled the doorway.

"Good morning," said the manager. Annoyance clouding his gray eyes did not match his attempt at a welcome. "The receptionist at the front desk will see you," he added to remove this interruption.

"I saw her," replied Caleb. "I'm here to see you." Caleb had a way of getting directly to the heart of anything he was saying. This piercing characteristic disarmed potential opponents while at the same time such clarity of intent removed impatience from the listeners leaving them ready to listen further with the realization there was interest available without any delay in dealing with this apparently old man who yet seemed youthful and presently carried a large basket. Caleb placed the basket on the floor before sitting down on a comfortable chair, facing the manager at his desk.

A woman entered the room. She wore glasses partially concealing blue eyes. Her short, blond hair was combed neatly. Her dress struggled to cover her generally round figure. She placed a cup of coffee on the manager's desk then walked out of the office.

"What do you want?" asked the manager without trying to mask his impatience. He sipped some coffee in the time he assumed he would have before this codger answered.

There was no answer. The manager had time, he calculated, for more coffee, so he sipped again and asked, "What do you want?"

Silence again and it was interrupting his coffee so the manager, to save his own enjoyment, asked, "Would you like some coffee?"

"Yes," Caleb replied.

"Cream or sugar?" inquired the manager.

"No," came the answer.

The manager walked briskly from the office and returned quickly carrying two paper cups filled with the steaming drink. He placed one on the desk in front of Caleb and saved the other for himself.

Picking up the cup, Caleb said, "Thank you. I'm Caleb Pine."

"I'm Harold Kirby," stated the manager. "What can we do for you today?"

"I've come here to open an account," replied Caleb. "I know you have a procedure for opening accounts at the front desk. However, there are complications involved with my situation and I don't want to attract a lot of attention to myself. I've been keeping some gold and jewelry in storage. I'm going to have to pay bills and don't want to pay with gold. Accordingly, I would like to exchange my gold and jewelry for money in an account."

"We can do that for you," stated Harold Kirby showing a first flash of interest.

Caleb started removing items from his basket, placing them

on the manager's desk. "This is a fortune!" exclaimed Harold. "I'm going to need some help," he added, standing and starting to walk toward the doorway.

"Not too much help because I don't like publicity," said Caleb.

"I understand," answered the manager before he rushed from the room.

Returning quickly, he said, "This might take some time. If you could leave your gold here, we will determine its exact value and put your investment in an account that I can open for you now."

"That's fine," replied Caleb.

"I'll do this work immediately," continued the manager.

"That's fine too," said Caleb.

After Harold had hastily prepared a new account then given Caleb a bank book along with checks and an advance of cash, Caleb left the bank and walked along Main Street until he reached a lawyer's office located in a building painted white with black trim. A bell rang when he opened a door to the office. He entered a spacious room and sat down in a chair facing a well-polished, oak desk. The room had an uncluttered appearance much like the man who walked through a back doorway. The lawyer's hair was combed neatly away from his face that had a carefully trimmed beard. Blue eyes peered through wire-framed glasses.

"Mat Holden," stated the lawyer, extending his hand toward Caleb.

"Caleb Pine," he replied, standing to shake hands then sat down again. The lawyer sat on a chair behind his oak desk.

"I bought land overlooking the Sand River," explained Caleb. "I would like you to look after that transaction and also take care of the purchase of some more land."

Caleb stood up, stepped toward the desk then placed on it a sheet of paper. Using a pencil to draw on the paper, he said, "This line represents Main Street, running in a north-south direction." Adding a second line parallel to the first, he said,

"The next street east is Westbury. Going northward, it crosses Church Street and goes to High Street." Marking more lines on the paper, he explained, "Proceeding southward, Westbury crosses the Sand River, cuts through brush country then forest and reaches Forest Drive. If we follow Forest Drive eastward, we come to Line Road. Heading north, up Line Road, we cross the Sand again then Church Street and come back to High Street. If we take High Street westward, we will cross Arrowhead Road then River Street and return to Westbury. Everything in that area, other than the section I already own, I would like you to buy on my behalf. The only part of this district I will not purchase will be the stone church and its piece of property on the north side of Church Street. My purchases will include the row of houses on the west side of Westbury along with the high-rise building being constructed east of Arrowhead Road. I already own the center chunk of land. I will just be expanding."

"That's a lot o' real estate," exclaimed Mat. His eyes clouded as he used his pencil to do some calculating. "With your pencil and paper, you've invaded much of the town, mainly its southeastern section. Are there limits set on the cost?"

"A reasonable price will be the best price," answered Caleb. "Buy the land as soon as possible. There is much to be done."

"Why do you want this property?" asked Mat with undisguised astonishment.

Caleb relaxed in his chair, noting sunlight shining through a window and brightening the room while Mat removed a thermos and paper cups from under his desk. Filling two cups with coffee, he said, "I usually keep this supply for later; however this will help us get through a long story that I seem to see coming."

After sipping the strong coffee, Caleb said, "In the past, the Iroquoians came here from the south. The site of this town is generally their northern limit for settlement. Farther north are the Algonkians such as the Ojibway. The Iroquoians once lived toward the setting sun near the mouth of the Mississippi River.

They were neighbors of the Wolf, or Pawnee Nation. The Iroquoians left this region and traveled up the Mississippi River to the Ohio River and reached the Great Lakes. One group, settling south of Georgian Bay, became known as the Hurons. Farther to the south, settled the Tobacco Nation then the Neutral. The Wenroe went southeast of the Neutral. Along the southern shore of Lake Erie, camped the Cat, or Erie Nation. Father south, traveled the Susquehanna, or Kanastoge, who stayed by the banks of the Susquehanna River. To the west of them, along the upper Ohio, went the Black Minqua. Up the Kanawha River, paddled the Nottaway and Meherrin. The Cherokee traveled southward across the Appalachian Mountains."

Caleb stopped talking to sip some coffee. He settled back into his chair and said, "The largest group of Iroquoians left Lake Ontario and traveled along the St. Lawrence River until they met Algonkians. Since these Algonkians flavored their food with bark, the Iroquoians called them Adirondacks, or bark eaters—porcupines. Battles were fought with the Adirondacks. The Iroquoians were outnumbered and defeated. They became subjects and had to pay tribute. The Iroquoians prayed to the Creator for freedom.

Silently, one night, the Iroquoians left and paddled up the St. Lawrence River. They went past the Thousand Islands and approached the Oswego River on the south shore of the beautiful lake, Lake Ontario. Here they were overtaken and attacked by the Adirondacks. In a storm, many Adirondack canoes were overturned and the remnant of their men returned home.

The Iroquoians proceeded to the banks of the Oswego River where they settled and planted corn, beans and squash. As time passed, the population increased and the people decided to move to new lands. Eastward, journeyed the Mohawks who settled along the Mohawk River. West of them, near Lake Oneida, went the Oneidas. Next were the Onondagas who stayed beside Onondaga Lake near present day Syracuse.

Farther westward, settled the Cayugas. The Senecas moved to the shores of Canandaigua Lake. The Tuscaroras traveled farther to the south."

"You're a long-winded fellow, Mr. Pine," said Mat, grinning.

"Thank you for that observation," countered Caleb with a smile brightening his face. "You'll be pleased to hear that the story continues. The groups I have mentioned became different nations. Battles were fought between themselves and also with the Algonkians until a peacemaker arrived who was called Deganawidah. He and his messenger, Hiawatha, formed peace between the Mohawk, Oneida, Onondaga, Cayuga and Seneca nations. They became known as the League of Five Nations later called the League of Six Nations when the Tuscarora joined after 1710. They are the People of the Longhouse, or Haudenosaunee."

Caleb sipped some more coffee and said, "The Iroquoians lived south of here. However, one village lived on the property I am purchasing. I am going to bring back the village."

"That's why you want to buy the land," exclaimed Mat.

"Yes," answered Caleb.

"I thought you might get to the point of your story if I waited patiently long enough and didn't run out of coffee," said Mat with his face brightening in a rare smile.

"You have done well," replied Caleb.

"I will look after the purchases and do the paperwork," said Mat.

"Thank you," stated Caleb, standing. "I appreciated your coffee and patience. Please work quickly. I must also get to work."

"I'll get on it right away," exclaimed Mat enthusiastically.

Caleb stepped out of the lawyer's office, left the building and enjoyed being greeted by sunlight of a new day. Wedges of light, flashing through mottled clouds along the horizon, sprayed crimson hues across fog drifting into town. Walking in this fog, with its patches of light, Caleb came to the office of the Clarksville Construction Company.

He opened the front door and entered a spacious office. Natural light was provided by windows in the north and west walls. The back, or east, wall was covered with notices pinned above a littered desk where a woman worked. Caleb stepped forward to a counter as the woman looked up from some papers. Her hair was grayish-brown and well curled. Her nose and chin noticeably protruded below dark eyes that observed Cal through gray-rimmed glasses. "You're lucky," she exclaimed. "I've been too busy to lock the door. However, we are open if we have a customer. Do you want to sit down?" She motioned to chairs beside the door.

"I won't be long," answered Caleb. He sat down in a straight-backed, wooded chair. "I'm Caleb Pine."

"I'm Rose Sims," she replied. "My husband, Slade, and I are the owners. I look after the office. He does the outside work."

"There is a lot of rubble," said Caleb, "on my property bordered by Westbury Street, Forest Drive, Line Road, and High Street. In the central part of my property there are old foundations along with walls of concrete or brick and some sections of asphalt. I would like to have all that debris removed in order to have the land look the way it once did."

"A project like that is wonderful news," she exclaimed while she wrote a note in a book. "We can begin the removal in three days."

"After the land has been restored to its natural state," continued Caleb, "I would like you to replant all the trees."

"Yes, we can do that," she stated almost in shock while she made more notes. "Will there be anything else?"

"After the land has been restored, I would like to have a traditional, Iroquoian village built. I will mark the locations of the longhouses and palisade. I would like a carpenter, plumber and electrician to finish my lean-too home beside the tall pine at the center of the property. The lean-too site will need a load of firewood."

"We know people we can get to build the traditional village," she almost whispered while her thoughts gathered all the implications of this wonderful new work coming to her construction company. "Did you know that Clarksville is located on the site of an old, Mohawk village?"

"Yes," he answered casually in his seemingly, always calm manner. "I'm going to rebuild the old village."

"There are Mohawk people around here," she said. "We will ask them to build the village structures."

"Thank you," said Caleb as he stood up and started to leave. "I'll look forward to seeing work crews arrive in three days."

As he opened the door, Rose said, "We have met before, haven't we?"

"Yes," he answered.

"Three days," said Rose before the door closed and Caleb stepped outside into a dense mist. He walked to a grocery store across Main Street and restocked his food supplies before heading south to the Sand River.

Fires had cleaned the river. Clear, spring water was visible again. The river can restore itself now, mused Caleb. The polluting factories have been removed. Grass will grow again along with the trees. The river's tributaries are supplying clean water instead of the previous chemicals and other waste products.

Beside a spring, which fed the Sand River, he stopped to rest for the night. At a fire's edge, he boiled water for green tea before roasting a trout that he had purchased at the store. Tea, along with the trout, provided a fine, evening meal.

After the meal, he relaxed beside the fire. He watched fog drifting among eerie forms of the night. An owl hooted occasionally. He watched the firelight create patterns in fog then he slept until dawn.

Morning sunlight crept along lingering patches of snow amid wisps of fog. Caleb used freshly ground beans to prepare coffee. He sipped it, enjoying the rich, earthy flavor. In the distance, a crow cawed. When a crow flew overhead, Caleb reached into his pack, removed a piece of bannock and threw it onto a patch of snow. The crow swooped down, picked up the bread then flew back in the direction of the hilltop pine. While mist swirled skyward, a robin sang, along with some purple finches. In the distance, the whistling call of a cardinal rang through the mist.

Back from the riverbank, Caleb cut down a maple sapling. He trimmed away its branches before peeling off the bark, leaving a straight shaft of preferred length. Next he held the stick over flames until it was hardened as well as mottled by burned areas. Pleased with his new walking stick, he used it to jump across a narrow part of the Sand River. He proceeded across burned land containing sections of brush.

I like using a walking stick, reflected Caleb. It helps when I'm climbing or jumping things and can be useful for checking strength of ice or depth of water. As a robin's melodious song was repeated from top branches of an elm, Caleb walked to the southeast. In soft earth, or patches of snow, he noted tracks left by deer along with rabbits, squirrels, raccoons, foxes and wolves. The weight of winter snow had pushed dead weeds into an even layer next to the earth. A flattened layer of weeds, as well as a burned section, revealed a square outline of an old cabin. A trapper's cabin, said Caleb to himself as his walking stick poked along the inside of the square. The stick dislodged two, clay pipes.

Maybe the trapper had placed these pipes on a shelf, or had hidden them some way inside a wall, reasoned Caleb while he held the pipes. The river is becoming clean again. Beaver and fish will come back. Wildlife and people will be able to drink the water. I would like to hear the trilling sound of spring peepers again. I have cleaned away much of the visible pollution although people are still poisoning the earth and water by misusing chemicals.

Forested, blue hills to the south had not been touched by fire. He walked to this forest and enjoyed its beauty and abundant life. Deer prints marked the earth along with wolf tracks. A ruffed grouse drummed announcing a familiar sound of spring. Ahead of Caleb a whir of wings followed the flight of a ruffed grouse. The rest of the flock exploded from a clump of hemlocks. A deer watched the passing visitor before feeding again on a patch of grass.

Caleb followed an old logging road that brought him to a ridge of land extending into a swamp. Mallard ducks flew overhead before circling down to a pond. On the ridge, Caleb saw an old, wooden door hinged to a tree. A second tree's trunk completed the doorframe located in front of a rusty, wood stove. Loggers used a tent to camp here while getting logs, observed Caleb. If logs are harvested carefully, the forest continues to live. These people were good workers. They did not use clear-cutting or chemical spraying and other practices like careless oil or gas removal and transporting that destroy the environment. We must have employment and an healthy environment. Too often the environment is wrecked to get short term and shortsighted employment. The loggers who worked here did no harm to the continuing life of the forest. They didn't ruin the wilderness. People have to harvest from the environment without killing it. A tree farm can be clear cut then replanted. If a forest is clear-cut, the forest is killed.

Caleb left the logging trail and observed a square foundation in the earth where there had been a settler's cabin. Rough mortar held fieldstones in place along the sides of a square cellar. Two birch trees grew from this enclosure along with a tangle of raspberry canes. Old apple trees grew near the foundation. By prints in the earth, Caleb noted that deer had been eating last season's rotting apples.

He walked farther into the swamp, following elevated stretches of land and crossing ponds by stepping along logs held in the grip of melting ice. The forest was often quiet except for a distant chatter of grackles. Occasionally a ruffed grouse drummed. To the southwest, the land was more elevated and undulated into a series of hills.

Using remnants of trails, Caleb picked his way through the forest. Above him, an eagle circled in a cloudless sky. With hills etched against the horizon on all sides, he arrived at a swamp where an old trail brought him to the edge of deep water. Two mallards flew up from the surface when Caleb stepped into water that was soon up to his waist. He waded past rotting trunks and places where ice continued to cover the surface. Keeping to the edge of iced areas, he proceeded to shallower water and reached an island.

He stepped into a creek then followed its watery trail trough a wall of cedars. Beyond these trees, the stream came to a falls. Beside it Cal climbed a rocky route to high ground topped by maples surrounding a meadow. Deer grazed in an extensive, grassy area. An hunter could always find food here if his family was hungry, reflected Caleb.

The deer looked up although showed no continuing concern with Caleb's presence. Walking beside the meadow, he crossed a narrow area between ponds then climbed a steep bank before entering an area of hemlocks. At the base of a rock ridge, a spring sent a stream of clear water to the ponds. In a deep cut of protecting rock, he saw a beautifully formed and designed pottery bowl.

On a bed of old ashes, he placed kindling topped by larger chunks of dry hardwood. Flames climbing along this wood soon flickered beneath a beefsteak skewered on a stick resting on forked stakes beside the fire. Heat waves shimmered above the flames, cooking the meat quickly while fat dripped into the fire.

Freshly crushed beans and spring water provided flavorful coffee. Caleb savored the steak and coffee and enjoyed the peacefulness of his surroundings. Along higher ridges, beech trees added their gray trunks to webbed patterns of the forest. Two mallards completed an half circle bringing them to a splashing stop on the adjacent pond. The fire flickered against a tangled backdrop of the woods and added warmth to other campsite comforts.

Dark forms of geese passed above the treetops as resonant honking calls rang through a mist-cloaked wilderness. After the meal, Caleb prepared green tea. He sipped the tea and rested while watching the surrounding landscape. The fire grew brighter with the advance of nighttime.

In the morning, Caleb left his camp and entered a mist filled woods. Deer watched him cross the meadow then return to the stream bringing him to the large pond. He waded into its cold water and hurried to the dry part of the trail. In his pack, he carried the Mohawk, pottery bowl.

Melt-water was gradually receding, leaving more bare ground and easier walking. Air was laden with fog. Quacking calls of ducks penetrated a rising veil of mist. Sap dripped from some maple branches. Geese honked from flocks, in flowing ribbons of life, streaming northward in response to a call of spring.

Caleb Pine emerged from the forest and proceeded toward fields located south of Sand River. Some of this section had been burned. Looking northward, he noted that the town blended with the landscape more than before the fire. The stacks had vanished. The only tall structure protruding against the sky was the high-rise building constructed east of Arrowhead Road.

Pleased with the way the environment was being restored, Caleb walked to the river. Clear water once again flowed along its usual course. Trout will come back, noted Caleb. Fish will return along with all other creatures.

He crossed the river at a shallow place where water splashed over a pebbly bed. A partially peeled poplar branch floated past. Picking up a second chewed stick, he thought, beavers have started to come back.

Caleb dropped the peeled stick into the water. Watching the whitened wood get carried away by the current, he thought, I feel a chill that doesn't come from the weather. I wonder what trouble is lurking ahead.

In a protective, rocky hollow in the riverbank, he hid the pottery bowl. Afterward, he walked westward. Near the bridge at Main Street, a muskrat crossed the stream. A snake slithered from the bank and entered the water.

Reaching the sidewalk bordering Main Street, Caleb proceeded northward. Again I feel a chill of impending trouble, he warned himself just before, in front of him, two men stepped out from an hotel doorway. One man's stringy hair hung below his hat and met a matted beard. The eyes were dull. The second man also wore an hat. He had a full beard. His eyes retreated in shadows under heavy eyebrows. Both men looked back through the open doorway before one shouted, "There's some fun outside. Come on out."

In response, the hotel spewed out a motley crowd onto the sidewalk in front of the old man with a walking stick. There was a cold edge to the air under a gray sky. The men came forward hastily.

"We'll just have some fun," shouted the man with stringy hair who had been the first to leave the building. His partner held a knife.

The old man's walking stick hit the talkative man in the crotch. He screamed, bent over then rolled onto the cement. The others pounced. The stick was swung with an unexpected swiftness that made the shaft difficult to see like a wind that crumpled the attackers. They fell to the sidewalk, yelling and holding broken bones until the old man was the only one left standing. Afterward, he resumed his walk northward. A few people peered from windows and doorways.

Caleb walked to the market where he purchased some groceries then returned to the riverbank to get the pottery bowl. From here, he went back to his lean-to.

He put pieces of bannock on the bird feeder before frying bacon in addition to extra bannock. After toasting bread, he sat down to enjoy a meal of bacon and tomato sandwiches. On slices of hot bannock, he spread butter followed by honey. Lastly, he sipped coffee slowly while sitting with his back leaning against the pine's trunk. A robin's melodious song rang through mist drifting gently across the landscape. Maybe I like this time of year the best, reflected Caleb. There is the spirit of a new beginning when the earth renews itself.

Above the lean-to and tall pine on the hill, sea gulls circled in a misty blue sky. Shafts of sunlight touched the bird's feathers with golden tints. Sounds of motors brought Caleb's attention to the arrival of a work crew. The workers have been busy cleaning up after the fires, he noted. There is much to be done on my property. The equipment is being parked for the night. One man is approaching my camp.

"You must be Slade Sims," said Caleb as the large man sat down beside the fire.

"I am," he answered, accepting a cup of coffee.

"You put anything in it?" asked Caleb.

"No and thanks," he replied. "You are Caleb Pine?"

"Yes, and you are on time," Caleb answered.

"I like to be," he said. He was a tall man with craggy features topped by graying hair. His blue eyes were those of a practical man who noted his surroundings carefully without making a show of anything. The serious looking face broke into a smile when he said, "I hope you won't be beating up any more of our town's citizens."

"I'll try not to," answered Caleb.

"You have given me a lot of work to do," said Slade before drinking some coffee. The cup seemed too small for his gnarled hands.

"I would like to have the land returned to the way things were before the asphalt and cement took the place of trees," he answered. "I have started uncovering a spring. Its water follows an old streambed to a foundation. The foundations and walls should be cleared away. The basement southwest of camp is filling with water and will become a pond. Its water will flow southeast to the old, stone bridge on Arrowhead Road before continuing to what used to be a great swamp where the tall building is now being constructed."

"I wondered why there was a bridge in such a dry place," noted Slade. "The construction in the area of your campsite, blocked a spring. You have released the water. Now it will flow again under the old bridge. I'm going to enjoy this project."

"When the concrete has been removed, as much as possible, more trees can be planted in the maple grove to the north, and a mixture of elms, pines and spruce trees can be placed southward," said Caleb. "After the land has been cleared and trees planted, the traditional village can be started." Giving Slade a map, Caleb continued to say, "This map will show the location of the longhouses and palisade."

Folding the paper and putting it into his pocket, Slade remarked, "You are well organized."

"I have more plans also," replied Caleb as he noted the darkening colors that had come with the evening. In contrast, the fire shone more brightly and Slade was pleased to receive a refill of coffee. The moon appeared from behind a moving bank of cloud. Moonlight lit the area, dropping pools of light beyond the crackling fire.

"We will get started early in the morning," said Slade. Returning the cup, he added, "Thank you for the coffee. I will be getting back." He stood up and started walking in the direction of River Street. Gradually his outline moved beyond an obscuring web of shadows. Caleb noted a shadow moving steadily across the landscape. He looked up in time to see a great horned owl flying toward the pine. Crows cawed and the owl veered away.

Caleb was pleased to rest in the lean-to. Moonlight became brighter. He watched the starlit sky then slept until dawn with its gray light giving increasingly distinct outlines to the area.

By the time sunlight brightened the ground beside the lean-to, Caleb had put fried bread out for the crows and he was savoring some of the hot bread himself. The morning was cool. Clouds moving in from the west gradually blocked the sunshine and brought a light rain.

Workers arrived early. Bulldozers and other heavy equipment moved forward onto the land. Trucks hauled away loads of debris to be dumped where clean landfill would be useful.

Picking his way past moving equipment, there appeared the form of Mat Holden. He sat down beside the fire and received a cup of coffee from Caleb.

Mat warmed his hands on the cup, sipped some particularly flavorful coffee and said, "You now own the property south to Forest Drive, east to Line Road, north to High Street and west to Westbury. Your land includes the row of houses of the west side of Westbury. Within this outlined area, there are just two properties that are not in your possession. A section on the north side of Church Street continues to belong to the church as you requested. The other section I was not able to buy is the triangular lot southeast of Arrowhead Road. A developer is constructing an high-rise building on this property. He will build others after the first has been completed. He refuses to sell the land."

"You have done well," said Caleb.

"I have papers for you to sign," replied Mat. He gave Cal a pen and the signing was done using Mat's briefcase as a desktop.

"After the land has been restored," said Caleb, "there will be a traditional village built here in addition to other buildings. There will be more legal work for you to do when you have time."

"Thank you," replied Mat, obviously pleased. "I have all the time you need. No one would realize you are as rich as you are."

"Money is useful," responded Caleb. "It is not an end in itself; but sure is an useful means to an end. There is much work to be done."

"Why don't you buy something for yourself—like a comfortable house?" asked Mat. "Your plans don't seem to include anything for you."

"Slade has electricians, plumbers and carpenters fixing up my home here," replied Caleb. "I have all I need."

"Your plans are good for this town," affirmed Mat. "We had become almost an abandoned place. You are putting us on the map again." Placing the cup on a flat stone beside the fire, he said, "Thank you for the coffee. I must get back to the office. As you say, there is much work to be done."

The lawyer walked away, following a path leading toward River Street. He greeted Slade Sims who was heading toward the lean-to. Taking Mat's place beside the fire, he was served coffee and also accepted a piece of fried bread topped with honey.

"Mat Holden is looking after the legal framework for our plans," said Caleb. Giving Slade a check, Caleb explained, "This has been prepared for you. "We are just getting started."

"And a wonderful start it is," exclaimed Slade smiling. He folded the payment and put it into his pocket.

"Mat has secured deeds to the land bordered by Forest Drive, Line Road, High Street and Westbury Street. This includes the houses on the west side of Westbury, and doesn't include the church land north of Church Street or the high-rise structure east of Arrowhead Road. All other burned and abandoned structures should be removed. Test for and safely eliminate any chemical of other poisonous contaminants. An orchard could be planted on both sides of the stone church. A mixed forest should be replanted on our land east of both Line and Arrowhead Roads. We need lots of trees. Along with cleaning the air, they bring birds, animals and beauty. After we have purchased the high-rise site, we will restore the swamp. The wetland and mixed forest east of Arrowhead will be fenced so we can keep our own deer herd. The herd will supply the village with food and hides—sort of like a cattle ranch. Many deer will be released each year to the surrounding forest. An healthy environment makes people healthy. I notice you have removed the visible part of the foundation where the spring water is flowing again. A fine pond is developing there. Please obtain some trout and put them in the pond as well as in the Sand River."

"You can be a long-winded fellow," noted Slade with a smile crossing his rugged face.

"I've been told that before," replied Caleb. "After we have returned the land back to its original beauty, there will also be, once again, a traditional village in the center of this area. To the north, there will be an orchard. In the east, there will be a deer herd. Circling from the east to the south, there will be fields for the growing of corn, beans, squash and sunflowers. Inside the stockade, up from the river, we should plant some tobacco for ceremonial use."

"With all these things to do," said Slade, "I Must get back to work." He stood up and placed his cup beside the other on a flat stone. "I have put a friend in charge of that village construction," added Slade. "Her name is Lily River. She and her husband, Callin, have worked with us for years. I talked to her about the village and she said she had already started getting materials for the site. She and her husband know the traditional customs. I thought they would enjoy this project as their ancestry is Mohawk and the village is Mohawk."

"I think you can work as fast as I can talk," replied Caleb.

"Now that's a real compliment," exclaimed Slade. "I have to get to work. We should soon have the land cleared and village started." He left the camp and walked southeast toward some distant bulldozers.

Slade keeps crews busy day and night, observed Caleb with satisfaction. He's a practical, hard working person.

From his hilltop camp, Caleb watched the equipment clean the land. He sipped green tea while fog stirred and shadows crept across the landscape with the approach of evening. Sea gulls cried from overhead as melodious songs of robins sang to the earth.

Caleb routinely fed the crows. Chipmunks also started coming to camp for peanuts. Near the lean-to, groundhogs dug a burrow. Both groundhogs enjoyed resting on a rock beside the burrow. From this location, they could watch the lean-to and receive select pieces of food.

Caleb was particularly interested in the flow of clear, cold spring water. It gradually filled a sand banked pond where the foundation had been. Having filled this pond, the stream continued along its course to the southeast. At Arrowhead Road, the water moved under the old, stone bridge. Beyond the road, the stream's route became lost while water drained into, and rejuvenated, a vast swamp extending to Sand River. Days became weeks and life returned.

After routinely checking the creek, Caleb returned to his camp. Slade Sims visited and introduced Lily River, a slim woman of medium height with short, black hair. Her almond-colored eyes sparkled with interest she seemed to constantly find in her surroundings, particularly the character, Caleb Pine.

Caleb served bannock topped by butter and honey, followed by coffee. Sitting beside the fire, the people watched moonlight brighten the landscape beyond a tall fire.

"The land has been cleared and trees planted," reported Slade. We used all our crews and completed work well ahead of schedule."

"In an old stump we were cutting for firewood," said Lily, "my husband found a group of flint arrowheads. We have known there was an old, Mohawk village site near Clarksville. However, when we cleared the ground to build a traditional village, we could not believe—and still can't believe—that we discovered outlines of an earlier village that correspond exactly to the plans you gave us. The earth is full of artifacts such as shards of pottery bowls and pottery pipes. There are awls, bone as well as flint arrowheads, grindstones, scrapers and adzes. We have been very careful while working in the area. We are rebuilding the village just as it was."

A shadow moved across the moonlit earth. Crows cawed before a great horned owl veered away and stopped at the next tallest, freshly planted tree. There was an interval when only snapping sounds from the fire broke the solitude before resonant hooting calls pierced the night.

"Caleb," stated Lily, looking at him directly, "the exact location of the village site is one thing; but when we prepared the palisade, according to your plan, we found the forms of an old palisade. The new longhouses are in the same location as previous structures. Running down the center aisle of the longhouses we are constructing, there is carbon and ash that have come from fire pits in the first homes. The old village is exactly where you told us to build the new one. We don't just have a traditional village, we have a rebuilt village."

"You have done well," replied Caleb. "The earth is restored. Trees have been replanted. Fields are ready for planting. The village has been excavated. All projects are on, or ahead of, schedule." After pouring more coffee for his friends and himself, he said, "Traditions are important. The best things in life don't change."

Using his walking stick, Cal cleared a section of sand beside the fire. While the end of the stick started outlining a map, he explained, "You are reconstructing an old village of five longhouses surrounded by a palisade. The palisade goes around this camp and then the pond and encircles the village. Please expand our new palisade to enclose our land along River Street and Church Street. Also construct a building in the shape of a longhouse that opens up along River Street. Back from this main structure, construct four more buildings in the design of a longhouse. These new structures will be museums that will display, as well as sell, traditional works such as pottery. We can use the sugar bush, north of us, and start a maple products business at the corner of River and Church Streets. The fields will soon be producing crops again. We will sell food. North of the stone church, please build an office and medical complex in the overall design of a turtle. North of the turtle, there will be a village with houses designed in the style of longhouses. The same type of village will be constructed west of us between River Street and Westbury. People who work in our new village can have the choice of also living there. There will be a deer compound east of Arrowhead Road and the pond along with the river will be restocked with trout. I've gone over these plans previously with Slade. The lawyer, Mat Holden, is looking after legal matters. I keep in contact with the mayor and council. They are excited to see their city and its region coming back to life rather than falling into decay. For political concerns, we'll need a council room in the turtle building. The first village that was located here traded over a large area. This trade will return."

"The old village will live again," exclaimed Lily. "Mohawk people and others too will work here. There will be full employment just like in earlier times." She received more coffee from Caleb then said, "There are ways of finding old villages. You can find a shard of pottery in a plowed field. Sometimes an arrowhead is found. That is how Arrowhead Road got its name. Occasionally a shard of pottery might be found in the mound of earth that a groundhog has removed from its burrow. Sumac trees like growing in ashes from old, cooking fires. Villages were always near a source of water—often a spring. Caleb Pine, you have found this village. You are excavating it and rebuilding it. The village of Clarksville is being reinvaded. The Indian nations were the first people to live on this land. Later came the Europeans. Now the first culture is returning. Everyone will benefit from such a return of this land's story. However, there is one thing—in particular—that I don't understand. Caleb, what method did you use to determine the exact location of the old village? How did you know where the palisade posts had been placed and where the longhouses had stood?"

Moonlight continued to spread across the landscape while flames flickered in the camp. An owl hooted repeatedly. "I can read signs that are on the land," said Caleb to his two friends. "I also remember things. I knew this village was a good place and traded with a large area."

"You certainly are bringing new life to the old village along with the town of Clarksville and its region," responded Lily, deciding not to push her questions any further at this time.

"This is certainly the best, largest and most wildly interesting job I've ever had," stated Slade. "With all this work to do, we should all maybe get some rest." He stood up and placed his coffee cup on the stone beside the fire. "Thank you for the coffee," he added before starting to leave.

"Thank you, Caleb," said Lily as her cup was also placed on

the stone. She walked with Slade finding the path clearly etched by moonlight.

At the campsite, Caleb enjoyed warmth from a tall fire. He watched the moon with its light and shadows. He slept until rain tapped on the lean-to next morning.

He observed the rainy day from the shelter of his camp. He used spring water to make coffee amid sounds of machinery moving upon the land. Work crews were active, both day and night, in varying types of weather. While days became weeks, Caleb's plans started to become buildings as the old village stirred again and came back to life.

Always eager to check the work's progress, he picked up his walking stick then proceeded to the spring. It had settled down to a steady flow of clear, cold water. A stream fed the pond that had become home to some mallard ducks. Geese claimed the eastern shore. The water's calm surface was rippled by these birds or broken by a flashing form of a trout.

Continuing eastward, Caleb came to new plantings of large trees including elms, white pines, white spruce and black spruce. Deer were both inside and outside of the compound across Arrowhead Road. Workers from the high-rise have been poaching deer, noted Cal. Each year some deer will be returned to the forest. A number of released deer will decide to remain around the compound and village. In earlier times, people could go to the forest and take what they needed. Today the environment's beleaguered resources have to be protected and even restocked. Hunting is an outdated activity because in modern times there is a shortage or scarcity of almost all types of wildlife including the forest itself. We must enjoy the beauty remaining in wildlife and not kill it.

Upon reaching the traditional village, Caleb was pleased to see so much progress being accomplished with the palisade and longhouses. From the top of a pole, a robin sang. Other robins sang in the distance.

Farther west, the village was being completed. This site was once part of a complex trading system for the exchange of

Indian products, recalled Caleb. European goods later followed the same routes. Indian goods are back in business again along with all items of trade. The museum being built here will send products along old and new trails. A fine product, like a good deed, can travel great distances and live through time.

Caleb proceeded to the Sand River where clear, clean water flowed again. The form of a trout darted to a deep pool. A muskrat swam across the stream, disturbing some wood from beaver cuttings. Red-winged, black birds chattered from willows at the water's edge. The earth is healing quickly, mused Caleb with satisfaction. Maybe I'll soon hear a whippoorwill's call.

From a bank of the Sand River, he went to River Street then walked north. On the east side of this street, the museum, offices and stores were being constructed. Along the west side, the village was being started with houses using the longhouse design.

At the corner of Church Street, the maple sugar operation was flourishing. To the northeast, an orchard had been expanded. North of Church Street, the turtle-shaped building was being constructed. Beyond it, there was a second, new village. The farming areas included land near High Street and stretched eastward to the southern forest. I must remember to purchase abandoned farms north of High Street and east of Line Road, noted Cal. Farmland is an essential part of the village.

Returning to the Sand River, Caleb followed its bank until he came to the old, industrial section west of Clarksville. Charred rubble had been removed and new buildings were being constructed. From one of these sites, the mayor approached Caleb.

"I've been wanting to see you," exclaimed the mayor. "I wanted to thank you for invading this town. You have put us back on the map. Instead of being an abandoned wasteland, the area is now busy. You brought us back to life. We are being reported in the news as a bustling community. Even the tourist business is starting because people now have reasons to come here. Consequently, new motels are going to be built west of town. For a welcome change, business is starting to boom along Main Street. We are not going to sell out to the polluters like we did before. We now know we have to protect the environment. We sold out the land in the past to make a quick, short-term, shortsighted profit. We have learned its better to make less money in the short run, and make sure the area doesn't die because of pollution in the long run. We did here in the past what they are doing now with the tar sands in Alberta and the pipelines along with fracking. They are rapidly exploiting the resources while destroying the environment for the present and the future. We need employment and an healthy environment. Everything depends on the environment. If the land is healthy, the people will be healthy and they also stay in business. I haven't felt this optimistic in years. I want to thank you for bringing life back to this town."

Looking toward the construction sites, the mayor said, "I was checking this new industrial area. We are making sure that this time there is none of the pollution that ruined us the last time. We aren't aiming at just reducing pollution. We don't want any of it. The town engineer, who is looking after these buildings, has been told that the water and air leaving here has to be clean. We have the technology to run clean businesses. We just can't let greed guide are profits. The only developer we haven't presently got under control is the one who is constructing the high-rise building east of Arrowhead Road. He has political connection, money—and greed. He will do anything immoral; but is careful to not do anything illegal. We are going to have to improve the laws in order to control him. Meanwhile, he will continue to be a problem."

"I'm watching him," replied Caleb. "All the people from whom I purchased land both wanted to sell and benefited from the arrangement. The only landowner I'm pressuring into selling is the one who is building the high-rise. The pressure has been as slow as the returning flow of water under the Arrowhead Bridge. The water will bring back to life a swamp where the high-rise is being built. One should never go against the forces of nature—especially the spiritual side."

"That's an interesting development that maybe no one but you has even thought about," exclaimed the mayor with a light of joy brightening his face. "The swamp will force the developer to sell?" he asked.

"Yes," answered Caleb. "Building in a swamp is not easy even when you know the swamp is there."

"That's the most fantastic and amazing news I've heard since I heard about you the last time," exclaimed the mayor with continued exuberance. "You've brought this whole area back to life. I must get back to work checking this construction site for pollution standards. Been great talking to you—and hearing about you."

Caleb returned to the riverbank and followed it to a secluded place where he built a campfire. From a pack, he withdrew a chicken he had purchased at the market. He prepared a skewer and placed the skewered meat above a steady flame. The skewer was held in place by forked stakes at the fire's edge. As the meat cooked, he sat down and rested his back against a pine trunk. He relaxed and watched the fire while occasionally turning the roasting meat. The area is returning to the way life was when the land was healthy, mused Caleb.

When the meat had been sufficiently roasted, he enjoyed a delicious meal while robins sang at dusk. When evening approached, an owl hooted before Caleb slept beside his fire.

In the morning, he walked to Main Street. Air was warm and humid under a clouded sky. New leaves on bushes added a greenish hue to the landscape.

He was enjoying the morning until he saw the men leaving the hotel. They first formed a group then walked steadily toward Caleb. I must remember to get a police force for the new village, he noted before tightening his grip on his walking stick.

While the men approached, Caleb recognized some individuals who had attacked him previously. The crowd was more hesitant this time. Ahead of the others, there stepped one man. He was a large, fat character who had tried to push his bulk into a tee—shirt, jeans and sandals. His greasy, black hair tangled with a beard revealing little of a face that was flat and fleshy with small, dark eyes. Carrying a baseball bat, he strutted menacingly forward.

Caleb's walking stick moved with a speed that made the shaft no more visible than a blur when it scraped across the man's toes. He screamed with pain and rolled on the sidewalk, holding his feet. The other men turned and hastened back to the hotel. Caleb picked up the bat then walked to the hotel doorway. When he stepped inside, his attackers, along with a few others, scrambled out a back door, leaving behind only one man who wore an apron. He walked behind a bar just before the baseball bat crushed the first table then the next until they had all been splintered. The bartender hid behind the bar to miss the first chair thrown to a back wall containing shelves lined with bottles. More chairs were hurled in a volley, knocking out mirrors, lights then windows.

"Bartender!" stated Caleb in a voice that was not particularly shouted although sharply heard. In response to the call reverberating through a heavy stillness left behind after the explosions in the room, the man stood up behind the bar. His face was ashen and his eyes bulged. Go to all the rooms in this joint and spread the word that there is a fire!"

"What fire?" asked the dazed man.

"This fire," answered Caleb before he lit a match, walked toward a window and lit a corner of a curtain. A flame spread along the material and flickered toward the ceiling. The bartender pulled a fire alarm then rushed up an adjacent stairway. After visiting rooms on a second floor, he hurried down the stairs and ran through the back doorway.

Having double-checked the building to make sure no people remained, and satisfied with the progress of the flames, Caleb left the building then continued walking along the sidewalk beside Main Street. We must have police for the new village, he reminded himself. A sparrow chirped from an hedge beside Caleb before police cars and fire trucks started rushing along the street. Wailing sounds of sirens filled the morning.

Arriving at the town library, he opened its side door then walked up a stairway to adjoining rooms filled with shelves of books. A scent of stale air seemed to fit a gray atmosphere of silence and stillness. A slim woman with gray hair and silver-rimmed glasses worked at a front desk amid piles of books. Looking up when Caleb walked to her desk, she asked, "Can I help you?"

"Do you have a section with books about this area?" he asked.

"We have lots of information," she said. "There is an entire section over there," she added, pointing to the closest aisle.

"Do you have books covering the Indian history?" he asked.

"No," she answered.

"Most libraries keep records of the past," said Caleb.

"Almost everything here is a record of the past," she countered.

"You have much information," continued Caleb, "although not the first records of Clarksville."

"That's were our local information is," she stated, pointing again, "and I have a lot of work to do."

"Yes, you have a lot to do in order to get a good record of this area," replied Caleb.

An icy stare was her only reply. Caleb turned away and walked to the stairs. Pleased to leave the confines of the library behind, he welcomed fresh air as he started walking along the sidewalk. A robin sang amid wedges of sunlight dropping from a mottled sky.

He proceeded northward and entered the lawyer's office. "I'll get some coffee," said Mat Holden. He left the room and returned carrying a thermos and cups. After filling the cups, he kept one while giving the other to Caleb who sat in the usual chair beside a window. Sunlight broke from a clouded sky, flashed through the window and marked the time of day with a bright streak across a wall behind Mat. He sat in his chair behind the polished desk.

"Thank you for the coffee," said Caleb after sipping some of the flavorful drink.

"I always look forward to seeing my best customer," affirmed Mat exuberantly.

"I came to thank you for the good—and fast—work you are doing in looking after the legal aspects of our village," explained Caleb, "and to get coffee too, of course. The legal side of any activity must be established first before any enterprise can be started. Law and order, along with the enforcement of good laws, are essential for the successful functioning of any activity including a business or community. People who are not protected by good laws are not free. Benjamin Franklin and Thomas Jefferson acknowledged their use of Iroquois principles of law when drawing up the beginnings of America's Federal Republic and Bill of Rights."

"There should be more known about the traditions of this land," observed Mat.

"I mentioned this need to the librarian," replied Caleb.

"I know there is nothing in the library about our early days," noted Mat.

"In regard to our present work," continued Caleb, "please look after the purchase of the abandoned farmland located north of High Street and east of Line Road. Also buy the forest south and north of Forest Drive. This abandoned farmland should be put back to work for growing food, particularly corn, beans and squash. We could also grow sunflowers, potatoes and tobacco. Tobacco is intended for ceremonies and not for personal use because this is unhealthy."

Finishing his coffee, and looking up toward the window with a serious, almost entranced, look crossing his face, Mat said, "Apparently unlike the librarian, I've been reading some records kept by my family. The founder of the present town of Clarksville, Hendrick Clark, set up a land company and made a fortune selling Indian land he did not legally own. The building for this land company was constructed on a side of an hill exactly where you now have your lean-to, or maybe slightly to its southern edge. I think your pond now covers the exact site of the old foundation. Knowing you, as I have had the privilege of doing, I've been tantalized by the incredible possibility that you selected this exact location for your camp because that was the site where the greedy scoundrel, Hendrick Clark stored his fortune that he stole from this land's first inhabitants."

"That's why I hired you," exclaimed Caleb, laughing. "You're far too smart to have on any side but mine. You remember that story about Spanish gold coming from South America? Well, the gold that came from South America wasn't Spanish gold at all. It was Indian gold that the Spanish stole and brought back to Spain. This stolen loot corrupted Spain and all of its settlements. Imagine how southern American countries would prosper if the Spanish could return what they stole—even just the gold. Well, coming back to the story of Clarksville, like so many other places, I know about Hendrick Clark and his land company's operation. He stole a fortune through fraudulent land sales and, as I have discovered, naturally stored it in his vault in the chamber under his land company's building." Standing, and speaking with his disarming blend of resonance and riveting clarity, "I'm returning the fortune! I'm giving it back to the founding people of this land and to the environment along with all people who respect the land and its history. That's what I'm doing. You're the first person I've told about the purpose of all these efforts. I usually don't stand up and make speeches to explain what I'm doing. I just do the work out of the pure joy and purpose of doing it. If people like me, they like me. If they dislike me, they dislike me. That's entirely their business—not mine."

"Holy jumpin' Jehoshaphat" yelled Mat. "I knew it and I'm kind o' proud of the fact I figured this out myself."

"Full credit to you," agreed Caleb. "Now you know why I wanted you to look after the legal side of bringing this area back to life. You're far too insightful to have on any one else's side but mine."

"Thanks Caleb," said Mat. "I'll purchase the additional land right away. Clarksville is going to prosper for the first time since the town was established—just like in the days when the first community was here."

"Now you've got the whole picture," affirmed Caleb before he stood up and put his empty cup on the desk. "Thank you for the coffee and good work—or maybe I should say good thinking. I'll get back to the lean-to and let you get your work done."

"Thanks for your confidence in me and giving me all this interesting work," exclaimed Mat. "I've never been so enthusiastic about anything before. Now, really for the first time, I can see the power and essential, overriding importance of legal work. I might sort of be saying that I'm seeing the real purpose of the kind of work I do."

"You have your life in line with your purpose for doing everything and such signposts, as this meeting today, that we meet along our journey provide excitement because they tell us we are on the right course and haven't lost our way," observed Caleb before he stepped toward the door and opened it. Before leaving, he said, "You are the first person to find out where I got the money to pay for the purchase of all this land—and the buildings."

"Yes," answered Mat. "The money comes from the forgotten vault under the long-vanished Clarksville Land Company that made a fortune by selling the same land—and others—that actually belonged to the people of the Mohawk village. All the details of this can remain buried under the pond."

"See you soon," said Caleb, "and thanks." He stepped onto the sidewalk and started walking back to his lean-to.

Slade and Lily arrived at the campsite shortly after Caleb's return. He served warm, maple syrup on fried bread accompanied by tea.

After lunch, Caleb showed his friends two effigy pipes he had formed from clay. At the point where the stem curved upward to become a bowl, there had been fashioned the form of a wolf.

Pointing to a bend in the Sand River, Caleb said, "At the river bend, the water is shallow and flows over a bed of the finest clay known to be excellent for making pottery bowls and pipes. This clay is of such fine quality there is no requirement of the addition of crushed stone for temper—or added strength. Tobacco should be used only for ceremonial purposes because smoking for personal use is extremely unhealthy."

With a sweep of his arm taking in the surrounding region, he continued to say, "Ash trees also grow well in this area. In surrounding forests there continue to be numerous ash trees that make the best baskets. The old village at this site was well known for pottery and basketry. Such items were used for trade. Our new village will revive these and all other industries that are natural to this region. There is no such thing as unemployment when the environment is healthy."

"Judging by artifacts we have found at the old site," added Lily, "the village was probably here about the year Seventeen Hundred. This year is approximately the time of the Iroquois League's farthest expansion—including northward to this site. Although the village moved south at a later time, the area around the Sand River continued to be the land of these people. Ojibway land is farther north. The Iroquois and Ojibway exchanged a wampum belt signifying mutual respect."

After adding wood to the fire, Slade said, "I'm going to stop listening to people around here because if I keep learning all this new information I could get so conceited I won't even want to talk to myself." A smile flashed across his rough features.

"Don't let that happen," laughed Lily, "because if you stop talking to yourself, we'll have to talk to you all the time—and we're too busy."

"I'll work on it," he countered.

After coffee had perked suitably, Caleb served it and said, "The old village produced much syrup as well as sugar. This custom and all related businesses will be revived. We will need a wild animal refuge with veterinarians to look after wild and domestic animals as well as birds."

With coffee, the friends talked beside the fire until a red moon appeared over the eastern horizon. As a red tint moved across the landscape, Slade said, "I should be betting back. Thanks for the coffee and hospitality."

"I should too," agreed Lily, standing and placing her cup beside the others on the rock. "We'll see you in the morning," she said to Caleb.

Slade and Lily left the firelight and entered the moon's red hue as they walked southwest toward a work crew.

Caleb slept soundly and woke up to be greeted by the first gray light of dawn. After rekindling the fire, he placed at its edge the blackened coffee pot. Following a breakfast of pancakes and coffee, he walked to Main Street to purchase additional supplies. He then headed south and crossed the Sand River. Although the land stretching in front of him was abandoned farmland, he could visualize fields of corn with innumerable leaves rustled by a breeze and glimmering in sunlight.

Continuing southward, he reached the forested hills. When he entered the forest, an hawk screamed from above treetops. The cries faded into a lasting solitude.

Caleb followed remnants of an old trail. In most places, there was little or no remaining trace of an early pathway; however, some sections were rutted deeply into the earth. Like a fox that is seen only briefly in a forest, Caleb was a shadowy traveler who did not disturb his surroundings. He walked past geese standing elegantly beside ponds. Ducks arced downward out of a gray sky to splash onto extensive ponds. Geese called from clouds overhead. A melodious song of a robin greeted Caleb when he came to a swamp encircling the island of the deer.

Caleb sat on the trunk of a fallen birch. Before him, swamp water stretched to a wall of cedars guarding the island. A slight indentation marked the outlet of a flow of a stream of cold, spring water. I enjoy the tranquility of this place, reflected Caleb while he observed the beauty of his surroundings.

Wedges of sunlight dropped through the forest canopy and mottled the area where Caleb rested. Not crossing to the island, but continuing beside it along an elevated area, he angled southward. The trail left the swamp then climbed into heavily forested, blue hills.

He stopped to watch five wolves walk down an hillside. Where there are wolves or eagles, there is wilderness, noted Caleb with satisfaction. From above the treetops, an hawk screamed. A web of light dropped in curtains of golden tints to arrange similar patterns on the forest's floor. A ribbon of splashing water dropped along a cliff near the trail. Caleb proceeded onward and upward until he came to a meadow that was part of a level area of brush covered fields. He walked to a thicket of sumac trees bordering the banks of a cold creek. Beyond fields of brush, there were blue hills.

He walked along the creek's bank until he entered a slight clearing amid a tangle of sumacs. Upon old ashes, in sandy soil, he built a campfire and rested while enjoying the warmth.

Using cooking supplies from his pack, he fried some bread and perked coffee prepared from freshly ground beans. A shard of pottery protruded from soft soil near the fire. A cheerful call of an ovenbird rang across the landscape while a few chickadees investigated sumac trees beside Caleb. Two of these birds accepted pieces of bread from Caleb's hand.

Following a restful lunch, he left his camp and walked toward blue hills in the east. Clouds drifted across a clearing sky bringing more sunlight to the land. The lone traveler continued following remnants of a trail opening to a ledge of rock. In a crevice, he located seven black, effigy pipes. Each one had the form of a wolf at the base of the bowl. He wrapped each pipe in a piece of leather before placing them in his jacket pocket.

From a particularly high outcropping of rock, he looked back at flat land to the west. He could also see surrounding blue hills. He rested on a rock that was well balanced and moved when he sat down.

He camped on the promontory and rested for many days. He liked to sit on the balanced rock and sip tea or coffee while watching the surrounding land. Things have certainly changed, he told himself. No longer do the passing flocks of passenger pigeons darken the sky. Some aspects of the past, like the passenger pigeons, won't be brought back and will only live in the spirit world, or the other side. However, much of the other goodness of this land's story will be maintained and told in many places including our new village.

Caleb Pine returned to his lean-to camp in Clarksville. The journey southward had relaxed him. He was ready to put new vigor into his work. He visited the museum to present its curator with the pottery bowl he had kept so carefully. He left this construction site and walked to the pond.

Upon his arrival at the pond, a wedge of sunlight left a break in moving clouds and swept across the water's surface in time to highlight a splashing trout. The fish flashed into a rainbow of spraying water.

Through new plantings of trees, Caleb walked eastward and checked the deer compound. From there, he continued on to the traditional village and was pleased to see that the longhouses had been completed.

In one of the longhouses, Caleb met Lily and her husband, Callin along with Slade Sims. Each person sat on a platform beside a fire in a center aisle. Callin was an energetic man of medium height and build with dark eyes and a face that was becoming lined from constant work he enjoyed doing. He served corn bread along with coffee. Afterward, Caleb removed some neatly wrapped leather bundles from his jacket pocket. A small bundle was given to each person. Carefully, the effigy pipes were unwrapped to be greeted by knowing smiles and appreciation. "They are a gift," said Caleb, "from the people whose story we are telling."

When the mayor appeared at the entranceway, he was invited to have corn bread and coffee. He sat down and savored the bread along with coffee. "I have good news," he said. "You can now buy the land where the high-rise development is located. You can therefore extend the deer compound to include much of the swamp as planned."

"Why is the land for sale?" asked Callin.

"Because," exclaimed the mayor while laughing, "the high-rise building is starting to sink."

"Why has it suddenly started to sink?" asked Callin.

The mayor looked at Caleb who replied, "When I unblocked the spring, the water returned to its old, natural course. The stream returned to flow again under the stone bridge at Arrowhead Road and brought back to life a vast swamp extending eastward between the bridge and Sand River. The workers, who started the high-rise, did not find rock to support the foundation; but a lack of rock is not unusual. Extra precautions are only taken if the land is thought to be unstable. This particular ground was thought to be suitable. The developer continued with the building. The quality or beauty of the land did not matter to him until now. Following the return of the water, the land became a swamp again with areas of very soft earth that would cause the building to sink."

Reaching into his pocket, and removing a leather-wrapped bundle, Caleb gave his gift to the mayor and said, "You have earned this."

The leather wrapping fell away quickly, leaving in the mayor's hand the beautiful, effigy pipe. "I'm always being astonished by you," the mayor exclaimed. "Thank you," he added, turning away. "I cherish this," he said before leaving the longhouse.

"We thank you for your pipes, Caleb," said Lily.

"It's truly appreciated," added Callin.

"I like being included too," said Slade.

"I thought you would appreciate these gifts," replied Caleb. "I must get back to work," he said before leaving the structure and starting to walk westward.

He visited the village site on the far side of River Street.

Pleased with the progress of the buildings, he moved on to the maple syrup center. He purchased a cupful of warm syrup. Sipping some of the golden-colored liquid, he looked beyond the sugar bush to see the initial forms of the turtle building.

Following the sidewalk beside Church Street, he walked to Main. He turned south and proceeded to the lawyer's office. Coffee was served before Mat sat in his chair behind the polished desk. Caleb relaxed in the usual chair beside the window. "The developer who was building east of Arrowhead Road is taking down that first building he started because he discovered it was sinking into that swamp I brought back to life," explained Caleb. "The developer has suddenly decided to sell this land. We want to buy it in order to extend the deer compound. Deer like swamps."

"I'll look after this transaction," replied Mat. "We've been able to purchase the abandoned farmland you requested along with much of the forest."

"We now have," continued Caleb, "a traditional village, a museum, two housing villages and a turtle-shaped office complex. We have replanted trees and brought back to life a trout pond, a deer swamp, a river and many acres of farmland. Associated with all this renewal, there are innumerable, connected business enterprises breathing new life and action to this whole region. I appreciate your good work in looking after the legal aspects of these projects."

"I have appreciated such a pile of challenging work," replied Mat. He served more coffee as sunlight lit the window and room.

Before Caleb left the office, he gave Mat one of the carefully wrapped pipes. "Thank you again for your work," added Caleb.

"I appreciate the work and this beautiful pipe," exclaimed Mat.

Outside again, Caleb walked through golden light from the setting sun. Sea gulls called from above the town. Caleb watched the gulls as they were etched in a last blaze of light from the sun. Amid cries of gulls increasing to a tumult of noise overhead, Cal thought, my work here has almost been completed.

Caleb reached the bank just before closing time. He hurried to the manager's office, catching Harold Kirby as he finished his notes. "I just brought you a present," said Caleb, "to thank you for helping me with our financial operations."

Always astonished by Caleb and now this package, Harold pulled away the leather wrapping to find the pipe, exclaiming, "I'm amazed by such beauty and to realize I actually know what it is—an incredible, black, effigy pipe. Thank you Caleb. I'll have a glass case made so I can display and take good care of your gift."

"See you soon," said Caleb before leaving the bank and continuing to walk along the sidewalk bordering Main Street. He turned east at Church Street and proceeded to the old, stone church. It looked out from the north side of the street and was bordered by the village's orchards. Farther north, there could be seen an outline of the new, turtle building.

"Caleb Pine," exclaimed the minister, Lester Jenkins, after he opened the church's door and saw his visitor. "Welcome to our church." Lester was a generally round-shaped man with a beaming and happy face.

The two men entered a stonewalled building containing rows of wooden benches. The men sat on chairs beside a front table. On this table, Lester placed fried bread along with mugs of tea. "I get the fried bread from the bakery in new village," he explained. "I seem to be there as much as I am here."

"We've all been busy—and seemingly getting busier all the time," noted Caleb.

"And thanks to you," replied the minister. "You certainly have been bringing new life to the town of Clarksville. You have our church surrounded—in a wonderful way."

"I have added water to the roots of an old wilderness and village. From this old village, a new one is growing. However, before traditions can be kept they have to be understood. When the early missionaries first arrived here, they did not all understand that the Indian people already knew and had much knowledge about the Creator who had formed the wilderness and wanted it protected and respected rather than just exploited as they were doing here and in other places today with ruthless oil extraction and fracking. The wilderness contains God's thoughts of beauty for this earth. To the early, Indian beliefs, the missionaries added the message of Christ who guides people to the Creator."

"It's amazing," exclaimed Lester, "that after we peel away all the misconceptions and misunderstandings between the early, Indian beliefs and the Christian message and get down to the core messages, we find that we don't disagree. We just keep learning."

Caleb sipped some good tea before replying, "Our new village could construct and help maintain a building at the back of your church to provide an additional place where people, requiring food or help, could come and get supplies. The fields around here will produce much food to be used for trade—for sale—or to just give to people."

"What took you so long to get here?" exclaimed the minister, beaming with extra happiness.

"Preparations," answered Caleb.

"How soon can we get started?" asked the minister.

"Lily River will visit you—most likely tomorrow," said Caleb. "You can decide about the construction."

"Once you get started, you move quickly," replied the minister.

"And I have something for you," added Caleb, reaching into his pocket. He withdrew a leather-wrapped bundle and gave it to the minister.

Unwrapping the present, he exclaimed, "One of those beautiful pipes. Thank you Caleb. Years ago I found one of these —just like this—at the base of a rock outcropping in the hills south of here. I'll show you the one I found." Hurrying to a side room, he returned holding a pipe in each hand. An astonished look crossed his face when he compared them, whispering,

"The pipes are the same. What is going on here Caleb?"

"They come from the old village," answered Caleb. "After being located here, the village moved southward. This site, with its spring, fine clay and marsh remained as an area for providing food and other supplies."

While Lester continued comparing the pipes, Caleb finished his cup of tea then stood up saying, "I must leave now."

Walking with his guest to the doorway, Lester said, "Thank you for your visit and the pipe."

"Thank you for your hospitality," replied Caleb. "Look forward to seeing you at the village," he added before stepping outside and starting to walk southward.

Caleb returned to his lean-to. He rekindled a fire and perked coffee before sitting down to watch the new village—and area—coming to life. I like to watch the work being done, he thought. Like any gardener, I must tend the growing crop. Too many people try to harvest nature's resources without planting anything. Oil, mining, logging and other such companies should not be allowed to tear apart the land—or water and air—for short term gain while leaving nothing but destruction for the future. Continuing employment comes only from an healthy, living environment.

Caleb walked to the pond. He threw pieces of bannock on the surface. Trout stirred the water as they fed on the bread. Next he went to the deer compound. He gave an apple to one of the deer that had become accustomed to accompanying him during walks through the swamp.

Proceeding northward, Caleb walked to the council room being constructed in the turtle building. As many people were present, Cal asked for a meeting to be held near the longhouses in the traditional village. They walked to the site, bringing in other people along the way.

A fire was started in a central area. Workers assembled along with visitors and they either sat on the grass or brought chairs near the fire. Smoke from the flames climbed a clear sky. An hawk circled near the trail of smoke. A robin sang when Caleb casually stepped next to the fire. "This place," he said with a voice that was not loud, yet—almost surprisingly—clearly heard, "is the site of an old village. We must keep what we have learned that is good from the past. Each land has its own story. It should be maintained and cherished. We must maintain this land's founding cultures. The land itself has to be looked after and respected. Freedom and equality have also to be safeguarded by laws. People have to know they are not intended to walk alone. If they do, it is by their own choice. People are individualized parts of the Creator and are intended to live with such company and come to earth for experience and development of their own choosing. Early Indian beliefs and the Christian message are pathways leading to the Creator."

Interrupting his talk to look at the surrounding landscape, Caleb said, "We can see here a new village. It is not a physical village that is the real message here. Physical forms bring attention to spirit life that lasts forever and the mysterious purpose of life is to be with the Creator."

As casually and surprisingly as Caleb had started to speak, he stopped talking and left the fireside to return to his lean-to. He put supplies in a splint, backpack then picked up his walking stick. Walking at his usual, steady pace, he went to the Sand River, crossed it and proceeded southward to the forest. He camped beside a pond. Ducks, stirring on the pond's calm surface, sent ripples under a mirrored picture of the surrounding forest on their way to the shore where Caleb kindled a small, yet steadily burning fire. A slight fragrance of smoke drifted about his camp. Crows cawed from a canopy of overhanging branches.

While camped beside the pond, Caleb visited remnants of old trails. He was completely at home in the forest. Along with his sparse meals, he prepared tea or coffee. Always, he watched the forest. He enjoyed the whir of wings announcing the passage overhead of ducks or geese—although there were no longer pigeons. Whippoorwills then owls called at night. Robins announced the mornings. From distant meadows, meadowlarks sang.

Caleb walked farther south and selected a stand of hemlocks for a new camp. Within the forest solitude, there was a constant gurgling sound of a nearby brook.

He placed a top pole for a lean-to across sturdy, hemlock branches. From this main support, secondary poles slanted to the ground before being topped by an overlapping layer of balsam boughs. A similar layer of boughs formed a combined sitting platform and bed.

After building a fire in front of the shelter, he boiled water and enjoyed a flavorful cup of green tea. A thin trail of smoke from the fire ascended through an overhead canopy of branches. Surrounding forest greenery was broken by lighter slashes of birch or beech trunks.

Caleb heated a slight coating of vegetable oil in a pan before adding a layer of dough and frying a thick slab of bannock. Next he unwrapped fish fillets he had purchased at a market in town. After being shaken in a bag containing corn meal, the fillets were fried. Bread, fillets and tea provided a fine meal.

Amid a gathering of night's shadows, firelight brightened the campsite. Caleb rested on his mattress of boughs. He slept until the gray light of dawn brightened the forest. After enjoying a meal of pancakes followed by coffee, Caleb started a return trip to the Sand River.

He watched a fox walking along the riverbank. The animal stopped to pounce on a mouse before continuing westward in view of the village. A gunshot, blasting from a building site, knocked the fox over the bank. A group of workers approached the river. They carried the animal back to the village.

Caleb walked to his lean-to as warming rays of sunlight broke from a clouded sky. He visited the village to check progress of work being done. He also bought squares of maple sugar at the sugar bush store.

Before returning to his lean-to, he asked for another meeting to be held in the open area among the longhouses. All the workers were expected to attend.

While people assembled, Caleb kindled a fire causing a line of smoke to climb the sky. To the gathered workers, he said, "I would like to thank you for your work in restoring this village in order that its message can be retained from the past. Good traditions must be maintained. A land's story should be told. Care for the environment should be a central part of our life. We use things from nature; but using must not mean destroying. Each bird, plant or tree has a purpose in the divine plan. The fox that was shot from the bank of the Sand River today was killed wastefully because of disrespect for life. Such things do not go unnoticed. There is much to be learned in order that the person who shot the fox will some day be as much a friend to the Creator as is the fox."

After speaking, Caleb returned to his lean-to. From here, he noted the growth of the village. Through the passing days and nights the buildings were completed. He found himself with less and less work to do. The affairs of the village were well managed by its council.

Caleb was increasingly found to be in the forest. He followed old trails and stopped at both new and old camping places.

After one particularly long journey in the southern forest, he returned, carrying on his shoulder, a long post. He peeled away the bark and put the post up in the open area of the traditional village. On the light colored wood, he wrote pictographs. While he worked, he helped Lily with the few symbols she did not know. She was able to read the record post and discovered that Caleb had written the history of the old village. "You can tell the story now," he said to Lily.

"The story has been recorded," she replied. "We must keep this land's founding culture."

Next morning, while sitting near his fire in front of the lean-to, he contentedly sipped coffee while watching remnant wisps of mist turn golden in morning sunlight. Again, he saw his friend approaching. The coyote was also caught in sunlight during her direct approach to Caleb's camp. Having grayish-silver fur and yellow eyes, she tumultuously greeted Caleb before receiving pieces of bannock and previously cooked trout fillets.

While the two companions sat in the lean-to, Caleb said, "You are late today. Usually you visit at night when others can't see you. My work here is complete. It is time for us to go home to Bear Trap Mountain."

# PART TWO

An hand lurched from the gutter and grasped the edge of a curb before the slush covered form of a man rolled to a warm grate on the sidewalk. A derelict breeze sprayed debris over the still form then continued to chase a paper cup along the nearly deserted street. Highlighting him, a patch of light dropped from the morning sun. Three pigeons flew to this section of the sidewalk. They walked near the man and rested in sunlight with its first touch of warmth in the new day.

Opening his eyes slowly because of the sunlight, Jedediah Speaker saw outlines of buildings reaching toward an opening where wisps of cloud drifted across an azure sky. A touch of ocher color from sunlight illuminated these drifting strands. This moving sky was like the fog clearing from Jedediah's mind. I hope I had a good time last night because such parties aren't fun the next morning—particularly this morning, he thought while he became more aware of his headache and the cold that was knifing through his body. I remember going to a party. Now I'm warmed slightly by sunlight and a grate on a sidewalk. There are more parties in Clarksville now that people are enjoying prosperity again for the first time in a long time. This region got generally what I need personally—a second chance. I need to start a journey—like the city did—to discover my true identity and eventual prosperity.

Recovering gradually, Jed noticed a speck circling in the visible patch of sky. He heard a distant shrill cry. I wonder if the eagle is calling me—but it can't be, mused Jed. I haven't thought about eagles since I was back at home in Heron Cove.

Sensing the presence of something—watching him, he turned and noticed the placid, yet alert, pigeons. Moving stiffly, he pulled himself to a sitting position with his back resting against a brick wall. Searching his pockets, he found a piece of biscuit or cake and shared it with the birds.

He stood up, feeling slush water trickle into his shoes. Crossing the street, he entered a restaurant he liked because it was clean with home-cooking type of food. He sat on a chair next to a back table offering a view of the restaurant as well as much of the street.

The waitress who approached him was a pleasant—looking girl with a slight blush of fresh tan complimenting her brown eyes and auburn hair brushed neatly back from her face. She recognized him from previous visits. His boots, jeans, plaid shirt and leather jacket were of good quality although soaked. He was of medium size and weight with straight, black hair, firm jaw and light brown eyes. He appeared strong, yet unthreatening. "I didn't realize it was so wet out," she said, eyes sparkling.

"It was where I was," he replied slowly. "The weather seems to vary wildly according to where you are at night and where you get up in the morning."

"I prefer a dryer climate," she replied, smiling. Removing a pad and pencil from an apron pocket, she asked, "What will you be having now?"

"Bacon, eggs over well and fried potatoes along with coffee," he replied. She was turning away when he continued to say, "Glass of tomato juice and bottle of steak sauce."

"That's to change the weather?" she asked, smiling brightly.

"Yes," he answered as a grin broke across his rough features. "In the last few minutes I've decided I have to move to a better place—a better world—my world."

"Well, welcome home," she exclaimed before turning quickly and walking toward the kitchen. She returned with a bottle of sauce along with a glass of tomato juice and a cup of coffee. "Does the juice and sauce really work?" she asked.

"Yes," he replied. "Helps to cure whatever bit you the night before."

She returned to the kitchen while Jed mixed sauce with juice until the mixture was thick and dark. He added salt and pepper before drinking the brew quickly. More slowly, he sipped clear coffee, feeling it start to clear his mind and focus his thoughts.

After waking up on a roadside, the time has come for me to make some changes, he thought. I've watched Clarksville and area come back to life—its true self. What the area has done generally, I must do personally and be myself—for a change. This goal will require a long journey—and an internal journey.

Watching more pigeons fly down to sit in sunlight near the first three birds, he decided to feed the birds after leaving the restaurant. He asked the waitress for a loaf of bread. After bringing the loaf to his table, she said, "I won't asked what it's for—although I would like to know."

"It's for the pigeons across the street," he replied.

"I didn't think of feeding them," she said. "I always have bread I could give them." She walked behind a counter then, carrying a package of bread, she left the restaurant and scattered food near the pigeons. Birds moved toward the food while she returned to the restaurant. Bringing more coffee for Jed, she said, "I have lots of leftovers I can give to birds behind this building."

The waitress returned to work at a counter while the warmth of more coffee stirred Jed's thoughts, bringing him again to the problem of waking up on a roadside. I must start to be myself, although not all of the past has been bad. I have enjoyed some of the comforts to be found in town—and friends. I liked working at the butcher shop, preparing meat and other food for people. All the time I've been here though, I've felt detached, separate—from everything—like I'm not a member. I haven't felt a sense of oneness, of belonging. I can see this place; but it doesn't see me. I must close this gap and find the work and path that is in harmony with my true path and purpose. Then every day will be a great day. The eagle reminded me of my hometown, Heron Cove. I know the Ojibway heritage of the Heron Cove community. However, I see such traditions at a distance as if they—like most other things—are not really part of myself. I am separate. I must stop being a stranger to my own life. I'm going to begin a great journey to follow that eagle home.

The waitress brought Jed a plate of steaming food. He finished the meal quickly then left a good tip before leaving the building and stepping into warm, morning sunshine. He felt a new urgency as if, for a long time, he would be in an hurry to get caught up on previous days and opportunities he had wasted on misdirected actions. He knew he was now on course that would take him where he was intended to be.

Touched by amber light from the sun, first flakes of snow drifted through crisp air. A moving veil gradually obscured the sun. In a muted rustle of snow, buildings became ghostlike. I've always been late for everything, Jed told himself as he rushed toward the butcher shop, especially my own life. He concluded his work at the shop then went to the bank to withdraw his savings. He purchased supplies on the way back to his small apartment. Before resting, he visited the building's manager and paid the rent.

When everything was ready for a new beginning, Jedediah crawled under a table at the back of the main room of the apartment. I don't think anyone, not even the owner, knows about this door, reflected Jed as he opened a small door leading to an attic. Using a flashlight, he removed traveling equipment he had saved since first moving to the town of Clarksville.

With his new sense of haste, if not outright urgency, he restocked his traveling packs and refurbished all other equipment. His supply was extensive yet he resolved to carry only most essential items. I'm going to have to leave behind most of my books, he thought. In the past I did most traveling by reading. I like Thoreau's books, and history particularly including American Indian traditions.

After preparing all his equipment, he secured packs on his back and left the apartment. He walked directly to the train station and soon was sitting comfortably in a train heading northwest. Looking out a window, he watched as the bustling town of Clarksville gradually became lost in the distance to be replaced by an increasingly remote wilderness.

When the train came to a village that by Jed's calculations was generally south of his home village of Heron Cove, he concluded his train trip and rented a room at a one-story motel with adjacent restaurant where he purchased extra food supplies.

The motel owner was a large, robust woman with closely cropped, black hair outlining a face with lively, brown eyes and white teeth. "We don't get many travelers here during the winter months," she explained while opening the door to Jed's room. "It'll be cold in here for a while. We don't heat rooms until they get occupied." Turning on an heater, she said, "This heater will warm the place in an hurry. Hot water will be available soon."

The room had a single bed along with a table and chairs with adjoining washroom. Windows were bordered by intricate patterns of frost. "Room looks great," said Jed to the woman before she stepped outside to return to the motel's office.

Jedediah placed his packs on the floor beside the bed then looked out the northwest window. Beyond icicles hanging from the roof there could be seen, in the distance, blue hills and mountains of a vast wilderness. I am drawn to this land. I want to know it and be part of it—like my ancestors understood it.

While the heater warmed the room, he stepped outside to observe the village. It consisted of a group of buildings scattered near the banks of a stream. In addition to the motel, there was one store, a garage, a few houses and a bar. Rarely, a car passed along the road. Smoke rising from chimneys marked a cloudless sky. I like this kind of community, he thought. It is still part of the natural environment.

He returned to the almost warm room and slept until the first gray, light of dawn. After preparing his supplies, he was eager to be traveling. With thoughts of the distant, blue mountains, he stepped out of the room, closing the door behind him.

A fox moved like a shadow down the street. This sleek animal was carrying a chicken. A dog barked. A raven called as large wings could be heard brushing against still, morning air. Snow squeaked under Jed's boots as he walked out of the village.

Before leaving the road, he put on his snowshoes. He commenced a steady pace across an icy crust topped by a fresh layer of snow. Because fox tracks were leading northward to the hills, much the same way he wanted to go, he decided to walk beside these prints that formed a line in the snow. Occasionally drops of blood appeared from the chicken.

The trail bordered an hedgerow before cutting across an open field. At the next hedgerow, the prints veered northward through the remnants of an abandoned orchard. A well—used deer trail proceeded north from the apple trees. Fox tracks ran parallel to this path and kept Jed following the fox. These prints led to an increasingly dense forest of poplar and ash with stands of cedar. Beyond high ridges between numerous ice and snow covered ponds, the terrain became rocky at the outset of the hills. The fox stayed away from a rustic house and came to cliffs looking out over a forest of spruce, pines and hemlocks. Rather than climb higher, Jed left the tracks and entered a vast area of swamp to the northwest.

Ice and snow covered ponds offered flat pathways between elevated ridges where aged willows sprawled branches and trunks. An eerie atmosphere stalked this swamp like a spiritual presence defined by its own remoteness.

Large, soft maples had fallen in some areas. When these trees fell, their roots followed the trunks upward carrying with them large circular upheavals of earth. In the shelter of a particularly large, upturned root system, Jed found a natural place for a camp. On a base of logs, he kindled a fire.

I've been thinking about roasting a chicken after seeing the fox carry one out of the village, mused Jed as he removed from his pack a butcher-wrapped chicken purchased at the shop where he had worked. Although initially frozen, the meat had thawed sufficiently to be placed on a skewer over a fire. While the meat cooked in heat above the flames, Jed sipped green tea and rested while enjoying eerily wild beauty of the surrounding swamp.

Something moved in the distance, to the southeast, at the edge of the adjacent snow and ice covered pond. Fascinated, Jed watched as the object advanced in line that, if continued, would lead directly to his camp. The form loomed larger and continued to be unidentifiable until Jed realized he was watching the sprawled outline of a chicken held in the jaws of the fox. The apparition came half way across the pond's surface before agilely turning around and retracing the same trail.

That fox is backtracking to mislead a pursuer, exclaimed Jed to himself as he watched the animal move out of sight beyond a tangle of willows where movement had first been seen. She later reappeared at the pond's edge, east of camp. The animal sat on a large, willow branch extending out slightly above the pond's surface. While that fox is eating the chicken, noted Jed, she is also watching her back trail. I too will keep watching the tracks and see what is following them. Like the fox, I will wait and dine on chicken.

The fox didn't have long to wait, thought Jed when he saw something moving along the trail. Approaching at a steady pace, there was an animal with tan and white fur. It's a large, strong dog, observed Jed. That critter looks as wild as the swamp. I'm not much concerned about wild animals because they are afraid of people. Dogs though are more of a potential threat because they don't fear humans. I heard a dog bark when I saw the fox in the village. If this is the same dog—and likely is —he doesn't give up. Uncasing his rifle, Jed thought, in town we call on legal authorities to handle many situations; but in the wilderness, I have to solve situations myself. Bones mark the places where people make their biggest mistakes.

The dog was preoccupied with the fresh tracks. When the prints ended abruptly on the open expanse of the snow and ice covered pond, the hunter looked up and saw Jed.

Dog and man looked at each other. That dog is of medium height yet has a wide chest and looks as strong and wild as this eerie and frozen swamp, Jed warned himself, keeping his rifle ready for action. We are equally surprised to see each other. I'll wait and let him come to me.

When the end of the dog's tail wagged slightly, Jed knew the standoff was over and he shouted, "Come on over, dog. Have some chicken. Everyone else seems to like it."

The large wild-looking dog stepped forward slowly and was encouraged by more friendly words that brought him into camp where he enjoyed pieces of chicken. The dog remained standing while eating. He looked over and saw the fox watching him. The fox then left the branch, moved out of view behind an upturned root and went farther into the swamp. The dog did not follow. "I suppose you've played this game before with the fox," said Jed to his new companion.

While the dog slept near the fire, Jed constructed a lean-to roofed with overlapping hemlock boughs. A resinous scented mattress of the same boughs covered the floor. Resting in this comfortable shelter, Jed worked at carving a bow from hard maple. When I was younger, he reflected, I liked to hunt with a bow and arrow. I always made my own equipment and always had a dog. This swamp and dog bring back memories of earlier days. I'll call the dog Bow.

While Bow seemed content to rest, or sleep, in camp, Jedediah completed the bow, followed by arrows then a toboggan. Satisfied with the results, and resting with a cup of tea, he thought, there are hazards to being here in the wilderness. A serious accident or injury could be fatal. Maybe I was wrong to leave the town; however, I needed something more meaningful than the life I had there. The town had returned to its roots and come back to life after years of stagnation and pollution. I had to do the same personally as the town had done generally. I likely should plan things more carefully. I have been too spontaneous. Maybe I just procrastinate until I have no choice but to follow a certain course. In future, I'll do the planning carefully first then the spontaneous parts won't disrupt the path to be taken. Well, I'm in the wilderness now, walking back to my home—my heritage. My Ojibway customs are both old and modern—like other cultures. As with the others, I don't have to live in the past to maintain my way of life. I do, though, find the roots of a culture to be as vital and interesting as the modern representations. There is danger in this swamp as there is in a town. However, I like the rugged, wild beauty of wilderness. There is an eerie stillness and a silence I can almost hear. I enjoy having the company of Bow. He is strong and wild like the swamp. He would sound the alarm if trouble approached, especially at night when I can't see as much of what is happening.

Jed left camp to collect firewood. Bow followed and seemed to not miss much in the forest world of scents, sights and sounds that were often out of reach of Jed's awareness. Jed found no human trails in the swamp other than his own. There were numerous deer prints and Bow located tracks of another dog.

Traveling farther than usual, Jed approached the southwestern edge of the swamp. The ancient tangle of forest gave way to rolling fields cut by hedgerows. Beside the closest hedgerow, there was in the snow a deep indentation of a snow machine track that swerved out of a grove of wild apple trees. At the side of this grove, a trail of blood stained deer prints circled back to the swamp. Following this trail, Bow and Jed came upon a young buck that had died from a bullet wound not long before they arrived. Jed brought the animal back to camp and saved both the meat and hide.

A chunk of venison was soon roasting over the fire. An aroma of cooking meat mixed with a fragrance of wood smoke and both scents drifted in crisp air.

I enjoy being in camp and sipping tea beside a fire, thought Jed while he sat in his lean-to and watched patterns of firelight move out toward tangled willows beside the pond. I find a sense of accomplishment in having made a comfortable place amid rough surroundings without doing any damage to the environment. The dog must have been a stray or had run into some mistreatment because he has decided to stay with me. I enjoy his company. He doesn't like to play like most dogs. He is content to camp with company and seems tireless when traveling in the forest. We make a good team.

Flames brightened above a deep bed of coals when fat dripped from roasting meat. At the edge of these coals, Jed placed the coffee pot. After a fine meal, both man and dog slept on a mattress of boughs.

When weather turned particularly cold, Jed pitched his tent and heated it by means of a small, metal wood stove with a pipe extending out through the roof. The stove kept the dwelling very warm. A floor of spruce boughs added a fresh, resinous scent to a fragrance of wood smoke. Bow enjoyed being in the tent.

During his ramblings, Jedediah noticed that the swamp was a refuge for wildlife. There were tracks of most of the regions wildlife such as squirrels, rabbits, porcupines, deer, foxes and wolves. Ruffed grouse and turkeys were also present. Occasionally gunshots from hunters or poachers blasted and echoed through the swamp. Blood stained deer prints were found along with a few deer that had died from their wounds. How many parts of a forest can be destroyed before the wilderness itself dies and we are left with a few city parks? Jed asked himself as he sipped coffee beside a fire in front of his tent. Every part of a forest that is lost is significant and every part saved is also vital. Law enforcement sometimes does not reach well into remote areas. Too often the law is on the side of a polluter or a shooter.

As a precaution for times of trouble, Jed taught Bow to seek cover when told to hide and then return at the sound of a whistle. The training became a routine that Bow mastered quickly.

While Jed was sitting in the lean-to, watching sunlight glimmering on drifting snowflakes, he noticed something moving on the far side of the ice and snow covered pond. That trail brings a lot of action, noted Jed while the object approached. It became more distinct until he recognized the fox. The fox is playing her old game with a new adversary, Jed thought. Fortunately Bow is asleep and won't interfere.

Near the center of the pond, the fox turned around and commenced backtracking along her old trail. Her form diminished and almost became obscured by drifting flakes of snow. She has stopped, observed Jed. He put down his coffee cup and watched with renewed interest. The fox started running back toward the center of the pond. "She didn't have time to backtrack," whispered Jed.

Bounding, with snow swirling away from her paws and sides, the fox ran toward the camp. Chasing the fox, and quickly overtaking her, was a large dog. Bow awoke in time to see the fox dash past in a streamlined blur of reddish fur. In headlong pursuit, out of a white haze of the swamp, crashed a large hound.

Bow slammed into the hound with such snarling fury that the hound was knocked over into a tumble of legs, fur, teeth and snow. Tearing away in a move that sprayed blood on the snow, the hound took a chance for retreat and put every muscle into racing back across the pond. Bow broke off the pursuit quickly and came back to camp, shaking his head, trying to dislodge a mouthful of fur. He and Jed both watched the fox leave the lookout point on the willow branch and return to the swamp.

I'm enjoying life here, reflected Jed while he filled his cup with coffee. For the first time in a long time, I'm aware of being on a course I'm supposed to be on with a purpose of following my own destiny, as if, being in this swamp is a guidepost indicating that I'm on the right track. I'm being myself and that brings a greater joy than just going through the sequence of the days of the week without being aware of where I was going and why. I feel closer to things here in the swamp than in town. I also think there is no such thing as coincidence. That word merely provides an easy explanation for an occurrence that indicates a prearrangement if we were aware of the road we should be traveling. I think meeting Bow is part of my journey that I will see more clearly later.

Jed added wood to the fire and roasted some venison. He and Bow rested in camp until nightfall. Intrigued by the beauty of moonlight glowing through the swamp, Jed decided to continue traveling to the northwest. Moonlight sent a network of shadows lurking across the snow. While the moon rose, shadows moved and Jedediah Speaker packed his equipment.

He walked with Bow to the low, aged, willow branch. On it he placed a frozen chicken he had purchased in the village. I've always liked foxes, he mused. If I could help them all, I would. All we can do is help those we meet and we do not know how far the ripples might go.

Returning to camp, he placed many supplies on his toboggan. He started traveling and enjoyed the steady whispering sound of his snowshoes moving through snow. He and Bow kept walking until they left the swamp and came to an area that had been a tall forest but had recently been logged. Stumps remained amid piles of brush.

Constantly checking tracks and scents, Bow crisscrossed through the route Jed traveled. At the base of a stump, Bow stopped to sniff cautiously at an indentation in the snow. The dog's careful approach alerted Jed who next noticed a few links of chain uncovered in the snow. Using his left arm to keep Bow from advancing farther, Jed used his right hand to get a stick and poke it into the indentation. In a spray of snow, steel jaws of a trap sprang up and clamped onto the stick. After a check with another stick revealed no further traps, Jed released Bow and he sniffed the now harmless enemy he would never forget. Stomping on the trap, Jed thought, animals and people are beset by dangers along the paths we follow. Trapping should never be done cruelly and only if there is an abundance of animals to be taken. A trapper should also need the income rather than just be adding on extra profit. What a waste, however to have Bow caught in a trap—or any creature.

Continuing their journey northward, the travelers came to a wide valley. Here a dark swath of open water colored the center of a broad river. Except for the center area, the river was covered by ice topped with snow. There is a bay in front of me. I could cut an hole through ice on the bay and catch some fish.

Walking to the river's edge, Jed used his axe to chop at the ice and found it to be suitably thick and strong. He ventured out onto the bay, stopping occasionally to check the ice.

With open water at a distance in front, and a breeze at his back, he first cleared snow from a patch of ice then chopped an hole suitable for fishing. Water, bubbling upward like a flow from a spring, quickly filled the opening. Ice chips were scooped away before Jed unwrapped a fishing line. He attached a silver colored lure then dropped it into the murky depths. He pulled the line at regular intervals to keep the lure flashing in circles.

The routine of jigging gave Jed an opportunity to observe his surroundings. Wind brushing against his back sent snow swirling toward open water. Clouds came from the west and darkened the evening. At night, the ice seemed to crack more loudly. The ice is cracking because of the increased cold of nighttime—and the wind, noted Jed. I did not intend to be here at night. We are tired, however, and need some rest. The ice is thick, although I don't like being near open water. Open expanses are normal along this river.

Bow's wanderings took him close to open water. Jed was about to call the dog when he veered away and started to come back quickly. After returning, he stood beside Jed, faced into the wind and growled. Jed turned around to watch the area leading toward the shore. He couldn't see much other than wind-driven snow. Hearing a muted pounding sound, he looked to his left and saw a large buck running away from a dog. That's the same dog that Bow fought in the swamp, noted Jed. I'm getting a cold feeling from this place. I should get out of here.

At the water's edge, the buck turned to fight. The dog attacked, pushing the deer back until both fell through the ice. Afterward nothing moved along the river except wind whipped snow.

Jed returned to fishing although he noticed the cold more that he had previously. The wind increased. Bow was restless. I'm getting out of here, exclaimed Jed to himself. Fish don't bite during a storm anyway. I have to find a more sheltered place. All signs indicate that it's time to leave.

When Jed started to bring in his line, it moved and slanted under the ice. He pulled the line sharply to set the hook in the biting fish. No fish could be felt on the line although it continued to slant sharply to the side of the hole. Water moved in the opening as if something was stirring in the murky depths. "The current has increased," exclaimed Jed to the nervous dog. "Why has the current changed?" he asked while staring into the cold night. "I hope the water is moving—and not the ice." As if in answer to Jed's fear, a loud, long cracking sound rumbled through the ice then he saw an expanse of black water widening between his fishing place and the shore.

Jed scrambled to pack his equipment. He put on snowshoes, thinking they would help to distribute his weight on the weakening ice. He and Bow both watched in fear as long, winding cracks broke away sections of ice. Pushed by the wind, large ice floes formed then diminished in size as they moved outward toward the far shore. The dark, center area of open water narrowed then vanished when ice crossed the river. A continuing sound of cracking ice pierced the inky night. For the first time in my life, thought Jed, I'm feeling the mouth-drying, heart-pounding sensation of raw fear. A rumbling crash followed the advance of an ice sheet over stationary ice along the north shore.

Amid a thundering frenzy of breaking ice Jed and Bow ran north. Maybe an axe or rifle will catch on ice if I fall through screamed Jed to himself while he ran with Bow. They jumped over breaks and slipped or fell when water surged across broken surfaces. Jed fought with an energized strength he had not felt before. He fell, scrambled and ran until he realized he was on unmoving ice. He and Bow kept running and fought across an upturned ridge of ice. They tumbled onto snow covered ground.

Both dog and man rested in a state of exhaustion. Jed enjoyed the security of having solid, snow covered ground against his back. He looked up at the familiar outline of Bow's head silhouetted against a paling sky. Scratching the underside of the dog's jaw, Jed said, "We made it. We have a second chance at life. I'm going to try to do more with my second chance than I did with my first. I'm going to try to enjoy life more, knowing who I am, following my intended, genuine course and thereby being of best help to others. If I'm lost, I can't guide others. As Davy Crockett said, 'Be sure you're right—then go ahead.'"

Fatigue overcame Jed and he slept while Bow watched the river. In the midst of exploding ice and thrashing water, Jed woke up screaming from his dream. Bow scrambled to his feet. He looked around at the shadows of night then stared at Jed. "Thanks for the concern," said Jed to the alert dog. "I'll be all right. We are going to have to be more careful out here in order to survive."

With his back resting on the snow, Jed scanned the starlit sky. Immediately above him, there loomed Bow's head as he guarded Jed against any additional trouble. Reaching up and scratching the base of the dog's jaw, Jed said, "I've acquired a friend along the course of my journey."

While the dog and man rested, dawn's light seeped into their surroundings erasing the night. "Light brings promise of a fine day," proclaimed Jed to Bow as they started walking again. "This is a day we almost didn't get to see so we'd better make the most of it—along with the rest of our days. I'm lucky that my equipment has not been lost or broken—except for the toboggan. I left it at the fishing hole."

Getting back easily into the rhythm of snowshoe walking, Jed headed northward. He and Bow came to a marsh laced with tracks of rabbits, squirrels, foxes and deer. Jed followed Bow along a blood stained deer trail until it came to a thick stand of cedars. Among the trees, water from a spring flowed into a pool rimmed by rocks. The stirring water prevented ice from forming over much of the pond. On a knoll, there was a deer that had recently died from a bullet wound. I don't like hunting, noted Jed. I would hunt only if I actually needed the food. I've been lucky to find fresh meat. Again I'll save the meat and hide. This is not deer hunting season. Poachers have been active.

Jed camped beside the spring and roasted venison for himself and Bow. Beyond the pool, the creek wound its way to a swamp. Brown grasses and bulrushes protruded above the snow amid a rugged tangle of willows. While watching wild forms of the landscape, Jed thought, like my Ojibway culture, most ways of life are old as well as modern. Citizens of European ancestry don't have to travel by covered wagons in order to keep their ways of life and I don't have to live in wigwams to keep mine. We keep our traditions in old and modern ways. The old Indian ways are kept while adding new information where it is useful—like other cultures. I will walk the road of the old ways and when I get back to Heron Cove I'll work to keep the heart of the traditions. I have always liked Indian designs and styles. In this place I'll build a wigwam.

Starting to work immediately, Jed tied poles near their tops and erected a frame to which he added more poles until he had a conical structure much like the start of a tipi. Around this frame, he tied a covering of overlapping sheets of birch bark stripped from large, dead, birch trees. Looking with satisfaction at his completed work, he thought, real birch bark coverings are peeled from living trees in the spring. Bark doesn't peel well in winter. I've made use of dead trees that were probably killed by acid rain.

A small fireplace was formed at the center of the floor. Smoke was released through openings between poles at the top of the lodge. Remaining floor space was covered with a layer of cedar boughs.

Bow liked his new home and was soon asleep on the comfortable boughs. Jed sat down on the bough mattress and enjoyed the small, central flame that added a slight scent of smoke to the fresh fragrance of cedar boughs. I like the form and beauty of this practical home, he said to himself. The wigwam springs from, and is part of, the natural environment.

Over a small fire outside the wigwam, Jed roasted more venison before making tea. The flames appeared to brighten while evening shadows crossed the snow. Soon the patterns of night were colored by moonlight. When clouds obscured the moon, only the paleness of snow provided outlines to forest forms.

A shadow of a great horned owl flickered across the snow as the large bird flew to cedars next to camp. The owl hooted repeatedly, filling the night with resonant calls. In the distance, wolves howled then silence returned broken only by a rustle of falling snow.

Jed went into the wigwam. He rekindled a fire while Bow continued to sleep soundly. Jed sat down and watched the movement of light along the inner walls of his lodge. Soon he was also asleep.

He was awakened at dawn by the sound of wind howling against the wigwam and trees. He quickly lit a fire to add its warmth to the lodge. He heated his hands above this welcome flame while Bow stirred at the back of the structure.

A strong northwest wind thrashed cedars and drove snow against the campsite. While streams of blowing snow reached across the sky, and whirled in occasional funnels, rays of dawn's sunlight broke from a bank of clouds and touched this moving snow with a spray of golden colors. Winding streamers of driven snow raced across the surface of the landscape.

Inside the shelter, Jed boiled venison and onions in a pot over the flame. A second pot boiled water for coffee. Following a meal of stew for himself and Bow, Jed leaned back against a support pole, and sipped coffee contentedly while observing the white world of swirling snow outside the entranceway.

By evening, the blizzard had ended. Stars once again sparkled above a quiet land. When the moon sent a path of light through the entranceway, Jed enjoyed the pattern of moonlight across a fresh layering of snow forming clear outlines of silver and shadow. He stepped out into this wild panorama. He and Bow left their camp and continued their northern journey.

Moonlight provided an intriguing brightness to the night. Gradually the moon turned to a pale glow veiled by a slowly moving haze of cloud. The night darkened while Jed came to a cleared section of an abandoned homestead. They walked to an open gateway in a wire fence. On the opposite side of this fence, there were dark outlines of apple trees.

Jed and Bow walked through the open gateway. To the north, beyond the trees, a field extended to a vague outline of a row of trees that appeared to border a road. The wire fence, that Jed and Bow crossed at the gate, stretched from west to east where it joined a similar fence extending from south to north.

Having stopped to observe his surroundings, Jed thought, I like the fresh, crisp cleanness of the land after it has been brushed with a new layer of snow. The wilderness gets more remote to the north. In this region, there are occasional places where people—pioneers—have attempted to establish farms. Some farms have been successful while many have failed because of a short growing season or poor, rocky land. Although there aren't many roads in this country, there appears to be a road ahead beyond the row of trees. There are many dangers facing me when I travel alone with Bow. However, I like the freedom to travel any place any time. Such liberty is a precious part of this trip.

Jed and Bow started to enter the field by the apple trees when high beam lights flashed from a truck parked beside the road. A swath of light crossed the field and highlighted a few apple trees. A second, moving beam of light searched the same area. These lights outlined forms of deer standing near the trees. Bullets blasted across the field as gunshots came from the vicinity of the truck. Three deer were knocked down to the right of Jed and Bow. Two of these animals stayed down while the third got up again and followed the other fleeing deer. Like bounding shadows with white tails flashing, the first animals easily jumped the fence then became swallowed up by the darkness beyond. The wounded deer followed the fence and hurried past Bow and Jed while more bullets screamed from the truck. Jed tumbled into the snow and held Bow down at the base of the fence. The deer eventually jumped it to follow the others.

"Before we get killed, I'll let the poachers in the truck know we're here," said Jed to Bow. Jed quickly uncased his rifle and aimed it carefully, squeezing off two shots, knocking out both of the truck's headlights. The other light was turned off.

An eerie silence gripped the night until the truck's motor started. The vehicle, without lights, moved slowly westward along the road.

Jed waited and listened until the sound of the motor became lost in the distance. "We should not stay here," said Jed to Bow. "They'll likely return soon."

While Bow wandered as usual, checking tracks, Jed walked southwest to another wire fence. They both crossed this fence at some high snowdrifts then kept to a southwesterly course across an hedgerow and onto an field containing an ice and snow covered pond.

This area is what I've been looking for, noted Jed. In case the poachers follow me, and they likely will, I'll use the fox's trick to hide my trail. Although I can't help Bow's tracks, he wanders so much only a real tracker could follow him.

Jed walked to the pond, leaving a clear, straight path of snowshoe prints. Near the center of the pond's expansive surface, he turned his feet around on the snowshoes then wore them backwards, retracing his own snowshoe prints and backtracking the way he had come. Upon reaching the open gate where the deer had been shot, he placed his feet on lower, horizontal wires and kept his hands on the top section in order to walk along the fence. A connecting fence took him northward to the road. Here his boot prints mingled with other prints.

If the poachers try to follow me, noted Jed, they will probably walk on my prints. This will make backtracking much more difficult after my trail has come to an end on the pond.

Bow's prints might be a problem; but he wanders so much the poachers might give up if they have to follow him very far. There are also other dog prints on this road. Although a good tracker could follow me, the fox's trick might just work.

Jed and Bow commenced walking along the road. They headed east then turned north at the junction of a second road. They returned to the forest by following a ski trail and soon left it to proceed west in first gray light of dawn.

I might be making a mistake, Jed warned himself. However, I'm going to walk directly south to see if I'm being followed. He hasted to high ground providing a panoramic view over the region where the deer had been shot. Two trucks were now parked beside the road. By means of binoculars, Jed watched four snow machines containing six people move out to the location of the deer. Two animals were loaded onto the snow machines before all four proceeded onward and stopped at the fence. Leaving the machines behind, the people climbed the fence then walked beyond the hedgerow to the ice and snow covered pond. At its center, the group stopped. After waiting for a short time, the people commenced walking and in single file returned to their snow machines. All four machines raced back to the trucks.

The fox's trick seems to have worked, exclaimed Jed to himself as he watched machines being loaded onto trucks. I'm pleased that I came back to see if I was going to be followed. The poachers haven't tried to relocate my trail. Their trucks are moving westward. I'll walk in that direction.

Keeping away from the road, Jed and Bow walked into a swampy forest of ice and snow covered ponds bordered by stands of poplars and cedars. Snowflakes started drifting across a cloudy sky as the travelers left this bush and continued over a field bringing them to an hill topped by a row of hardwoods bordered by an old, cedar rail fence. Jed and Bow walked along the north side of this fence and came to a maple grove with dense underbrush. This elevated position offered an almost unobstructed view of the area. Northward, there was a field with a forest beyond. To the east, stretched a mix of bush and field. In the south, a brush covered field led to the road. The region beyond this road included the place where the deer had been shot.

Continuing to find his binoculars to be useful, Jed scanned the area westward. Here the road met a second one running south and north. Next to this road three farms were visible. At the northern farm, someone was splitting firewood. The middle farm contained an house with a shed and barn. Two trucks were parked beside the barn. Those are the vehicles used to poach the deer, recalled Jed, returning his binoculars to a pack. This area is clearly a place where deer come for the winter and a lot of them are getting shot. Six people were using snow machines. That poaching business must be a large operation. Maybe the neighbors don't like what is being done. I'll talk to the person chopping wood.

While Jed walked toward the sounds of an axe splitting wood, he thought, I should just leave and mind my own business. But the destruction of the natural environment is each person's business. Destruction comes in many forms such as the Alberta tar sands along with fracking. In Clarksville, there was widespread pollution and everyone benefited from a return to healthy land and water. Poachers here are killing deer and likely other wildlife. I know what it's like to be on the receiving end of poaching as I almost got shot along with the deer. I can at least try to find out who almost killed me.

At Jed and Bow's approach, the woman leaned against the handle of her axe. Although wary of any strangers, she was intrigued by this man and dog. They were an interesting match in having a touch of wildness yet were unthreatening.

"Good morning," said Jed.

"I hope you're right," she stated, continuing her assessment of her visitors. The man was of medium build and height, yet portrayed the strength of a person who would be at home in the wilderness. He carried too much equipment to be part of the crew from the adjacent farm.

"The morning has not gone well for some deer I saw get shot down the road from here," stated Jed. He was intrigued by the contradiction the woman presented. Her gray hair could not camouflage an apparent lifestyle based on youthfulness and strength.

"The deer could've met some people from the neighboring farm," explained the woman. "The owners moved to Clarksville and have leased the property to their previous handyman, Neil Crew. By the way, who are you?"

"I'm Jed Speaker," he replied. "I'm from Heron Cove, to the north. I'm walking home. My partner is Bow. We almost got shot by your poaching neighbors when they killed two deer last night down the road by the swamp."

"That sounds like Neil and his friends," said the woman, making no attempt to conceal information. "They're a consistent bunch. They'll do anything that doesn't require hard, honest work to make a living. We, and most of the neighbors, have tried to stop what goes on over there and still the poaching and other things continue. The owners moved to town because of deteriorating health. Neil is supposed to look after the place. He does some trapping and deals mainly in meat and furs. He travels a lot and seems to be well connected in that business. Three men and two women work for him. The work is done in the main barn. The trouble in Clarksville has been cleaned up. Now we need to remove a few remaining pockets of trouble—such as the trouble across the road."

"I might have a look at their operation," said Jed. "Thanks for the information."

"What I've told you is not a secret," she said. "We keep hoping—and expecting—something will be done about it. I release some of my frustrations by splitting wood."

"Thanks again for your help," said Jed before he turned away and started walking back the way he had come. Stopping, he shouted back to the woman, "I'm going to have another look at your neighbors before heading home."

"Be careful over there," she warned before swinging her axe

at an upright piece of wood.

Jed returned to the hill where he could watch the poachers' farm without being seen himself. With his back resting against a cedar rail, he welcomed a chance to relax. His view covered a vast area. In the sky overhead an hawk circled slowly as if there was nothing of interest stirring on the snow covered land. Moving toward the hawk, clouds gradually came out of the west, bringing flakes of snow. Wind also gusted from the same direction. Soon, tendrils of snow whirled away from sharply edged drifts bordering the rail fence. A blizzard shook trees and filled the sky with billowing sheets of snow.

With a snowdrift at his back providing protection from the wind, Jed roasted venison over a fire then boiled water for tea. After the land seemed to shake itself free of the storm, Jed continued to camp by the fence for a few days to observe the farm.

"Well Bow," said Jed to the alert dog, "six people are working at the site. Two trucks transport snow machines to get deer or other animals. They are taken into the main barn. Customers drive to the barn to buy animal and likely other illegal products."

At night, light could be seen shining through spaces between boards on the barn—the center of activity. After lights had been turned off, six people walked back to the house. Vehicles were parked in front of a shed beside the house. The farm is a destructive place and it's one of many, reflected Jed, while enjoying a cup of coffee during his observations. Clarksville was a polluted place; but its return to a clean environment has returned prosperity to all people rather than just producing profit for a few polluters. The environment has often been transformed from a magnificent wilderness to a wasteland in a greedy rush to tear money from the earth. Fracking, as started in some places, poisons the ground and its water in a rush to make short-term money. Keeping an healthy environment is each person's responsibility because everything is connected. Something in me has awakened. It all started with the rebirth of

Clarksville. Now the same is happening to me. The deer, like all other parts of the environment, are part of—and for—all of us; and are not just for poachers. Poachers come in different forms. Some are oil companies. Others are mines. We need energy but only clean energy. Too often we hear that destroying land, water or air is just good business and protesters are radicals. I'm going to work to reverse this cycle to the place where polluters are know to be the radicals and environmentalists are seen to be good business people.

Pouring the last of the coffee into his cup, Jed concluded, the large number of deer here does not mean that they are numerous. They come from surrounding areas to herd in one place during winter months. The blizzard will have erased all tracks. I could leave now and not bother anyone. Moreover, no one would be bothering me. I don't have to stir the hornets' nest. I will not, however, walk around a situation where my help is required. First I'll help to stop the poaching then resume traveling north. I must be ready to act when the lights go out in the barn.

The next night, following a late meal, Jed and Bow left their camp. Before them stretched a gray, snow covered landscape mottled by a blend of forest and field. Above this panorama shone the still, clear circle of a full moon. A shadow flickering across the snow marked the flight of an hunting owl. With silent wings, it flew past the moon and later killed a crow that had been sleeping on a white spruce branch. A raucous clamor of cawing broke the night's silence while a flock of crows pestered the owl. Crows have a good reason for bothering owls, noted Jed as he approached a hedgerow he would follow southward. He and Bow crossed the road east of, and close to, the poachers' farm. The hedgerow provided cover, allowing Jed to get sufficiently close to the back of the barn. Lines of light appeared through spaces between boards in the old building. Dark outlines of the house and shed loomed against the lighter sky.

Jed waited patiently until the streaks of light blinked out, leaving the barn in darkness. Four workers, followed almost immediately by two others, left the building by way of a side door and walked to the house. Afterward, only lights from the house shone into the night.

Concealed by a drift of snow, Jed lit a candle. Its flame ignited the first of the flaming arrows. The arrows were pulled to full draw in the bow. When released, each shaft whooshed upward, sending streaks of light arcing toward the barn. The first arrow stuck just below the roof. Three additional arrows followed the first. One of the last three vanished through an upper, ventilation window. Flames spread slowly, yet steadily among the old, white pine boards. Outside flames spread then joined an interior blaze to increase in intensity until gathering itself into a roaring, twisting inferno brightening the gray sky.

Looking back occasionally to check the fire's progress, Jed followed an hedgerow leading east to the apple trees where the deer had been shot. A fireball lit the western horizon.

After walking through the gateway in the fence, Jed retraced his trail to the pond then backtracked as before to the fence. With his feet on lower, horizontal wires, and hands on the top, he repeated his walk along this fence. It joined a second fence bringing him to the road. It took him eastward. Rather than turning at the first connecting road, as he had previously, he walked farther east before turning north along fresh snow machine tracks following a bush road through an hardwood forest. He soon left this trail behind and pushed onward selecting his own route, heading northwest.

Jed prepared a camp in a dense stand of hemlocks. He roasted venison for himself and Bow then watched the glow of firelight gradually fade from a clouded sky in the southwest.

The next morning, by the first, gray light of dawn, Jed and Bow entered a vast swamp. Ice and snow covered waterways provided paths through a tangled maze of brown, snow topped bulrushes and willows. The rising sun colored the swamp with red light. Shades of red brightened to gold as the sun moved above willows and entered a cloudless sky. This sunlight marked the beginning of a winter thaw.

Jed prepared a camp amid fallen willows and waited for colder weather that would make traveling easier. Meanwhile, the landscape gradually darkened with water from melting snow and ice. Clumps of snow dropped from branches as winter temporarily released its grip. Water trickled to the main waterways. While the land absorbed warmth, water flowed.

Beside his camp, Jed watched the tip of an hemlock branch where a drop of water was touched by sunlight and also mirrored forest colors. From this observation, Jed noticed similar drops of light glimmering throughout the forest.

A growl from Bow brought Jed's attention back to camp and a worry that some danger lurked where only Bow could detect it. "I've learned to always pay attention to your warnings," said Jed to the agitated dog. "What's out there?" asked Jed when the growling resumed. "Maybe the fox's trick doesn't work twice and Neil Crew is following us." Jed knew that worry had crept into his thoughts and ruined his enjoyment of the camping place.

He packed his equipment and walked northwest. Progress was slow and difficult because of wet snow. After jumping a stream, Jed found himself looking into the pale eyes of a black wolf. There was a long patch of white fur on her right, front leg. Steel jaws of a trap held the other front leg. Although Bow showed no sign of attacking the wolf, Jed tied Bow to a tree. "Just being careful," said Jed to the dog. He seemed to welcome an opportunity to rest.

Jed cut the ends off a forked branch to make a long restraining pole. Pointing it toward the wolf, she pulled away as far as the trapped leg would allow her to stretch. The forked ends held down her neck while Jed used his boot and other hand to remove the trap. Continuing to be immobilized by a fear of humans, the wolf had to be probed with the stick before exploding into action and bounding away, leaving behind a few falling flakes of kicked up snow.

Before releasing Bow, Jed checked the area and found three more traps. He also removed a piece of meat that had been hanging from a branch above the trap that had caught the wolf. Having sprung all four traps, he reopened the jaws of two and set them under a covering of snow. On top of each, he placed a trap with its jaws closed. Neil and his crew use traps to poach animals, thought Jed. When the poacher tries to pick up the exposed traps, the two hidden traps will spring into action. Although this won't cause any serious harm to a person, it'll be a reminder of what a trap does. Trapping is a good line of work if done with care and the income is actually needed. However, Neil and his workers are just butchering the forest. We are all poachers until we feel the pain of others.

Jed untied Bow then they proceeded northward again. Leaving the swamp, they came to an hardwood forest laced with ponds and connecting flows of water from the melting snow. There will be springs here, thought Jed, as his snowshoes left a watery trail across a pond topped by ice and wet snow. Where there are springs, the moving water will not freeze.

He heard a sharp cracking of ice before he tumbled into murky water. Not knowing how deep the water would be, he thrashed wildly until he could scramble back to thick ice. He pulled himself then his equipment out of the grips of dark water smelling of rotting vegetation.

Chilling tendrils of water moved through his clothes and along his skin. To regain warmth, he hastened onward until they came to a region of fields sectioned by hedgerows and wire fences. Low areas were filled with water. All snow was watery. He followed an hedgerow to the top of an hill then came down again to flat farmland. The thaw continued with unusually warm sunlight.

After passing the edge of a small lake, Jed and Bow entered a narrow section of forest. A booming whir of wings marked the departure of two ruffed grouse. Colored like the forest, they could be seen distinctly only because they flew across a bright, evening sky.

"We will camp in this strand of forest," said Jed to his companion. "We need a place to rest as well as a chance to get clean and warm." Jed built a small fire and savored its warmth although the narrow forest provided little shelter from the wind. Bow seemed uneasy as if he didn't like the place. "Is it this location or something else that bothers you?" asked Jed who was concerned by the dog's restlessness. "If you don't like this place we'll leave. But first I have to rest." Jed sat close to the warming fire. Its flame sent light flickering into the murky woods. An owl hooted then Jed and Bow slept near the dwindling flame.

Shattering the grayness of dawn, the first sunlight painted clouds with red hues and cast similar shades across Jed and Bow's small forest. As the sun climbed the sky, warmth increased bringing more thawing and trickling flows of water. Jed roasted venison and watched overhead clouds turn golden above thawing fields. "We welcome a thaw for its warmth," said Jed to his companion who continued to be agitated. "Walking, though, will be more difficult."

While resting beside his fire, Jed watched a pair of crows land in a field to the north. This section had been plowed deeply in the autumn. The field was extensive, containing numerous hills. Snow had melted away from these hills, leaving large, deep furrows open to more warming sunlight. In the lower sections, water gathered. There is very soft earth on the hills, noted Jed as he refilled his cup with coffee. Water is seeping into the low areas. Bow is restless. He is worried about something. Maybe we are being followed because the fox's trick did not work the second time with Neil Crew and his gang. They are approaching and that's what is bothering Bow. Traveling across this watery land will be difficult, although there continues to be snow along the hedgerows.

Two crows flew from the south and joined the first two in the field northward. The four birds flew to the top of an oak, the highest point in an hedgerow bordering the edge of this field. The crows cawed repeatedly. "What's bothering the crows?" Jed asked the nervous dog. "Is it the same thing that's bothering you?" Trouble could be on its way, thought Jed. I'll get ready to travel.

While resting after packing his equipment, Jed noticed the eagle. It soared overhead in a wide circle passing through a golden haze of sunlight. There is always beauty to be found in the countryside, or wilderness, he reflected, and there is beauty here, even in this place that bothers Bow. Bow is too restless and he keeps looking to the south.

Jed shared some venison with Bow. After refilling his coffee cup, Jed said to his impatient companion, "We have been here long enough. Maybe the land will freeze tonight then we will leave unless trouble comes sooner from the south."

Jed noticed Bow watching a fox. The animal was running northward along the western hedgerow. "You don't often see foxes running in the daytime unless something is chasing them," said Jed. "We must leave tonight."

Jed carried a cup of coffee and walked with Bow to the southern edge of their narrow forest. Sitting down at the base of an oak, Jed wrapped his hands around the warm cup and sipped the stimulating drink, savoring its rich flavor. Looking south, he watched the small, ice and slush covered lake in addition to undulating fields and hills extending southward to lose themselves in a misty haze. Nothing stirred along the lake or toward the far, blue horizon. The land was motionless as if basking in sunshine. I take a long time deciding to do anything, trying to find the right course, reflected Jed. My journey has led me to this countryside in a winter thaw with enemies likely following me from the south. "You keep watching our back trail," he said to the agitated dog. "Don't worry. The night will be colder and we will leave this place."

When Bow growled, Jed looked southward and soon saw them. Four snow machines were keeping to remaining snow beside hedgerows and were approaching. "I hoped we wouldn't get caught here," said Jed to his nervous companion. "Fortunately, you know what is happening. I could tell you to hide then call you back with a whistle. However, you likely should stay with me."

Wearing snowshoes, Jed worked his way toward the center of the vast field stretching northward from their camp. Bow followed closely. Low areas of the ploughed field had become ponds while hilltops had turned to deeply rutted stretches of soft earth.

The four snow machines had increasing difficulty in reaching the trees. Earth in fields had become mud traps for snow machines. Even the snow was watery.

After checking the deserted campsite, Neal Crew drove his snow machine north to the edge of a very large, hilly field. Snowshoe tracks led toward the center of this field. Three other snow machines moved close to Neil's machine. Motors were turned off. Four people removed their helmets. "It's too hot to drive machines in this thaw," said Lou Dawson, the closest driver. He wiped sweat from his whiskered face. The other two drivers, Len Stickler and Burl Katel, were in a similarly uncomfortable state.

"Of course it's too hot," replied Neil sourly. His fleshy face was almost hidden by black hair and a beard. His dark eyes were glossy with a customary hostility that seemed to make him sweat more than the others. "This fellow we are following just might have burned us out of business. He also shot at us and caught my hand in a trap. Now he's playing the same ol' game. He has gone out into an open area, like before, and will backtrack, thinking that we will have messed up the tracks and won't be able to find where they continue. We'll circle this field and find his tracks. I'd like to know who he's working for."

Waving his arm eastward, he said, "Lou will go east around the field. Drive on snow beside the hedgerow. Len and Burl will go west. Circle until you find the trail. When you find it, stop and we will all meet at that place. We haven't seen any tracks branching away from behind us. I'm going to wait here to rest my hand. It aches badly after being caught in that trap."

The three machines progressed along their allotted routes while Neil watched from the edge of the trees. Clouds moved past the sun, bringing a chilling grayness to the day. The thaw is ending, observed Neil as he held his aching hand. Maybe everything is ending. I can't pay salaries other than mine. When the others discover I haven't any money, they'll leave. I hope I can keep my job at the farm.

The snow machines met at the far side of the field then slowly returned to the starting point where Neil waited. After the motors had been turned off, Lou, animated by frustration, exclaimed, "There are no tracks. Like the other times, we've lost the trail somehow. He has backtracked. We found his prints the last time because of the dog prints. Now there's nothing to follow. Maybe we'll see his branching off tracks on our way back."

"It's too wet and muddy here to be finding tracks," said Neil. "I would've liked to give that guy a present for all the trouble he has caused me. If he likes this land so much, we could have helped him by stickin' him into it."

"I think I'll stick to killing animals," said Burl.

"That business will have to wait for a while," replied Neil. "Right now I can't pay salaries."

"I guess you just forgot to mention that earlier," stated Lou.

The snow machines returned to the south. The sun shone again through a veil of slowly moving clouds. Caught in sunlight, snowflakes shimmered down to the land. Near the center of the field, close to the hilltop, deep furrows had been farther rutted by erosion. The earth was soft, becoming thick ooze in places. Along a deep furrow, an hand moved then a rifle. Jed allowed Bow to stand. Next Jed pulled himself to a standing position, trying to shake out the aches and cold. He secured his equipment and struggled northward across the field.

A moving curtain of snowflakes obscured the setting sun. As the temperature became much colder, the land's surface not only froze but also became snow covered. A deepening layer hid their back trail while Jed and Bow entered a white forest. Walking returned warmth to Jed, enabling him to wash in a stream beside a tall campfire. The night was calm and quiet except for the rustle of snowflakes.

In the first light of dawn, Jed and Bow followed the stream's bank through the forest until they came to a valley. Here the banks of a creek deepened on its way to an ice and snow covered lake. Before reaching the lake, Jed selected a camping place beside the creek near a beaver dam. A current of water flowing from the dam kept ice from forming on a long section of the stream. A campfire was soon flickering light into an adjacent forest. The night brought a clear, starlit sky. After cooking venison and preparing tea, Jed thought, there is a feeling of calm and quiet seclusion in this place. The poachers have gone away. My back trail is also covered with a deep layer of snow. After covering myself with mud, I must really take time to do some serious cleaning. So far, I have managed to keep most of my equipment except the toboggan.

Next to the fire, Jed tied together saplings and branches to form a dome-shaped sweat lodge that he covered with a tarpaulin. Using two sticks, he carried hot stones from the fire and put them in a pit located in the center of his lodge. After removing his clothes, he sat down inside the structure. When he splashed water onto the hot stones, steam billowed upward, bringing warmth and sweat. He savored this warmth and relaxed both his muscles and mind.

Leaving his steam-filled lodge, he walked slowly into the clear water of the stream. Sand as well as pebbles covered the creak bed. He welcomed the chance to get completely clean using fine sand as a scrubbing agent. The cold water also clarified his mind, seeming to leave behind few remnants of fatigue and worry.

I feel at home here, mused Jed after he put on clean, dry clothes. I'm getting accustomed to being in the forest again although I know I have much to learn. I'm also short of food. At night, I have noticed that light brightens low-lying clouds to the east. Such light must be coming from the village of Eaglewood. I know I'm north of Clarksville and east of Sky Lake. I'll have to go to Eaglewood for supplies.

While the rising sun shot wedges of light through the forest, Jed and Bow left their camp and crossed the stream at the beaver dam. They followed the stream's tributary because its banks led eastward. The first obstacle was a gorge cutting through the side of a valley. Jed and Bow worked their way upward beside water falling in tendrils through curtains of icicles. Sunshine illuminating this ice colored it and the tumbling water with rainbow hues. At the top of this gorge, the stirring current kept the creek from becoming completely covered by ice.

A few days of traveling brought Jed and Bow close to the village. The forest gave way to some cleared fields with fences and hedgerows. Light from the setting sun emblazoned flashes of gold on windows of a scattering of buildings.

Having the village in sight, Jed camped on the last outreach of forest. Following the loss of sunlight, the evening turned crisply cold. Some branches froze and split with thundering crashes that reverberated throughout the forest. A falling star flamed across the sky then stillness became as complete as the cold. A dog barked from the direction of the village.

The next day, Jed and Bow visited the one, general store. The building was spacious, containing general stock including bags of seeds emitting a pleasant, earthy fragrance to the air. "You look like you've been traveling forever," said a storekeeper who had a pugnacious appearance with watery blue eyes, uncombed hair and a protruding mustache.

"I feel like I've been traveling forever," replied Jed, starting to look around for supplies.

"Where are you headed?" asked the man for whom his store was a community and not just a business.

"I'm going north to Heron Cove," said Jed.

"You must like walkin'," continued the man.

"I do," said Jed, trying to shorten the talk so he could get supplies.

"You travelin' alone?" pushed the storekeeper.

"My dog, Bow, is outside," said Jed.

"Some critter will have you for dinner," stated the guy. "There are a lot o' wolves around and, in some places, you might be the only dinner out there."

"I might dine on a few critters myself," countered Jed.

"Don't let me keep you from spendin' your money," said the storekeeper, knowing he had probably pried out as much information as he was going to get.

"I'll have a good look around," replied Jed, pleased to be able to concentrate on what he needed. He quickly picked out a variety of necessary supplies, including dog cookies, finishing with the question, "Could we also get two, vanilla ice cream cones?"

With an incredulous look, the man said, "We don't get much demand for that at this time o' year. Two o' them?"

"Yes," he stated. "Blueberry, if you have it."

A woman, who had been busy near a back hallway, lifted the door of a freezer and started scooping ice cream from a tub.

"My wife gets the cones," said the man. "I'll get back to work. Good luck with your walk. Are you prospectin'?"

"No," answered Jed.

"A lot o' people are you know," stated the storekeeper as he finished checking the items. "The bill's out here," he said to his approaching wife. "I better get back to work," he added, walking to a back room.

Jed packed his supplies. He received the ice cream cones, holding them in one hand while he paid the woman. She had a round shape. Her face was lined with kindness and bordered by straight, white, closely cropped hair. She wore glasses and had pale blue eyes. "You're not a prospector?" she asked.

"No," he replied. "Just walking home."

"My husband's a prospector," she stated. "He's the only one left. This place used to be a mining camp for the Cold Creek Mine. He finds a little gold. I can't get him to leave."

"Thank you for these supplies," he said before leaving the building. Bow rushed to greet him, looking at the ice cream with excited anticipation Jed had not seen previously. Grasping the ice cream and cone in his jaws he snapped at the treat, quickly consuming it. Jed had time to only taste his treat before Bow was in front of him so eagerly expecting the second cone that Jed could only give it to him. I knew dogs liked ice cream, thought Jed. Bow, though, has a special craving for the stuff.

Jed and Bow left the building behind and commenced walking northward toward a distant blue landscape. Jed stopped once and gave Bow a snack of round, colored, dog cookies. Jed sampled a few himself, finding them to have a cereal taste and very dry.

The travelers were both pleased to get back to the forest. The snow was marked by tracks of rabbits along with an occasional fox trail and some wolf prints. Jed and Bow approached the top of an hill while the setting sun tinted a wild landscape with crimson hues. Crows cawed. Two gray wolves stood on the hilltop beneath a darkening sky. One of the wolves approached Jed and Bow. Bow growled as the large, black animal drew closer. "What's that animal doing?" said Jed to his battle-ready companion. "There's a long patch of white fur on the right, front leg. That's the wolf we took out of the trap."

The large, black animal stopped, raised its head and howled. The second wolf answered with a similarly long, resonant, piercing call. Agilely the lead animal swung around and both wolves raced quickly out of view.

The night air was calm. Jed did not need his tent. The campfire provided sufficient warmth for comfort. From a starlit sky, the moon shone pale light across the snow. A wolf howled in the distance. Silence returned, broken only by snapping sounds from the fire. A flame flickered above a deep bed of coals while Jed and Bow slept.

By the first gray hues of dawn's light, a slight edging of frost coated forest branches. This frosting glimmered in golden light from the rising sun. After rising through clouds, the sun sprayed shafts of color onto the snow covered land.

After leaving their camp, Jed and Bow walked to an ice and snow covered lake. Jed chopped an hole through the ice in order to try fishing. Wedges of sunlight continued to drop from breaks in clouds as Jed fished. Shadows of evening crept upon the ice, finding Jed without any fish.

He withdrew the line and was pleased to start traveling again northward to a region he had known when he was a boy. "We are returning to a place I have known," he said to Bow, giving them both the company of spoken words. "I used to hunt in the distant hills when I was younger, although I feel as young now. Like the earlier days, I am again living more just one day at a time. I can enjoy the renewal of each day with the rising sun and then reflect on the day beside a campfire at night. Maybe, as the elders say, all things do form a circle. I wanted to walk to Heron Cove in order to become part of the land again. The journey is longer than I thought and so far has taught that I know less than I thought I knew. Learning never comes without risk. We could vanish out here. No one really knows where we are."

Jed and Bow climbed to high country with snow-covered rocks and tall spruce. There were tracks of ruffed grouse, deer and wolves. Springs were numerous. They created a swamp mottled by patches of old, brown bulrushes. While leaving this swamp, Jed's attention was caught by the beauty of a small spruce emblazoned by sunlight. I'm seeing more things—becoming more aware—all the time, he noted.

The travelers reached a lake that Jed recognized from his youth. This ice and snow covered expanse brought back to Jed a flood of memories from old hunting, trapping and fishing trips. Like his memories, an eagle appeared. It called as it came out of a stack of clouds and entered an opening of azure blue sky. I will see, reflected Jed, if I can gather up the memories and save them to share with other people. To the always-alert dog, Jed said, "I know the ice on this lake is safe, and the fishing is good. We must catch some fish. Although I have never done this before, maybe we could just camp on the ice. The ice is not as sheltered as the forest. However, we could do more fishing if we could fish in our tent."

Jed put up his tent near the western shore to be protected from the prevailing wind. The shelter was tied to logs. They also provided a base for the small metal stove. Remaining floor space was covered with spruce boughs except for a fishing hole chopped through the ice.

When the camp had been prepared, Jed sat inside his tent and enjoyed a cup of tea steeped in a pot on the stove. The ice is an unsheltered place, he reflected. That explains why I've never heard of anyone camping on the ice before. Of course a tent camp is not much different from a fish hut. Water sometimes comes up around the huts. A person is wise to camp on shore. However, I have to catch some fish. This place is presently comfortable with a mixed fragrance of spruce boughs and wood smoke.

Equipped with supplies purchased at the store, Jed placed a pan on the stove then fried some bacon. Using the resulting oil, he fried three eggs. This meal was served on a paper plate. A steak was fried for Bow. The two companions enjoyed a fine meal in their comfortable, warm tent. Afterward, Jed and Bow slept on the layer of boughs.

Jed woke up in the cold, darkness of night. He rekindled a fire in the stove then lit a candle for extra light. Next he unwrapped his fishing line, attached a silver lure and dropped it into the hole. Golden colored water darkened over the descending bait. As the fire supplied warmth to this shelter and a candle added light, Jed jigged until a strong, struggling force hit the bait. When line was pulled downwards, Jed released line keeping it snug and gradually brought the fighter to the hole. Toothed jaws appeared at the opening in the ice. Water sprayed into the tent when a large, golden-colored walleye flashed out of the lake to be tossed outside the tent.

Jed's hand shook with excitement as he dropped the lure again into the water. A candle burned evenly, providing amber light while he resumed the calm routine of jigging.

Bow growled viciously and sprang out through the tent's doorway. A fury of snarling followed. Bow was forced back into the tent. With white teeth snapping from powerful jaws, a wolverine pursued Bow who crouched behind the stove where Jed had tumbled. A thrust from the wolverine's paw dislodged the pipe and knocked over the stove filling the tent with sparks and smoke. Jaws snapped and threatened while the furious attacker backed slowly out of the tent. Following the creature's departure, Bow and Jed scrambled outside into crisp, cold air.

Re-entering the tent, Jed set up the stove, reattached the pipe and extinguished a flame flickering among spruce boughs. Going back outside, Jed noted that the wolverine had vanished and had taken the fish. "Wrestling with a wolverine in a tent is not a good idea," Jed said to Bow who was sniffing tracks in the snow. "Any injury in the wilderness can be serious. Getting eaten can be a problem too."

Jed returned to fishing while Bow slept on the fragrant boughs. Dawn was casting a gray light upon the landscape when a pileated woodpecker called from the shoreline. Ravens croaked loudly and clearly.

Jed enjoyed the morning sunlight as he walked to shore to get firewood. While he gathered pieces of hardwood, Bow checked tracks particularly some well-worn porcupine paths winding between tree trunks. In a swampy area, there was a beaver lodge. A dam had raised the water level and killed a stand of hardwoods. With such an abundance of available wood, Jed decided to build a campfire. Flames caught quickly on dry wood, adding more warmth to a sunlit day. The swamp had a crispness and cleanness that gave Jed a feeling of exuberance. This forest seems to be celebrating something, he mused. Overhead, an eagle circled. Feathers shone in rays from the sun.

Returning to his tent, Jed kindled a fire in the stove. Warmed by this heat, and relaxing in the comfortable tent, he continued the routine of jigging until something strong struck the lure and pulled out line. A brisk struggle followed until a flashing form of a trout came out of the hole. Sides of the fish shone with rainbow hues. More efforts at jigging brought additional trout splashing through the hole in the ice.

In afternoon sunlight, Jed filleted most of the fish. He covered some in snow outside the tent. Four were carried to the campfire on the shore. They were barbecued over coals. This roasted meat was divided, in large steaming chunks, between Bow and Jed. Following a fine meal, they rested and watched evening shadows lengthen toward the tent on the snow-covered ice.

One particularly long area of shadows came to life in the form of a wolverine. It left the western shoreline and headed for the tent. Jed held Bow while they both watched their tent get raided. The wolverine uncovered a trout from snow next to the tent then returned to the western shore. The raider came back again before night settled upon a still landscape. "That camp robber will keep returning as long as food is available," said Jed to the watchful dog. "We must leave."

Jed and Bow walked to their tent. All equipment was hastily packed. The journey north was soon continued. The two travelers crossed the lake illuminated in fitful moonlight dropped from a clouded sky. Pushing onward, at a steady pace, they climbed to high, rocky country topped by lofty white pines. Within this forest, shallow gravel-bottomed streams murmured between banks covered by ice and snow. The water flowed clearly and sparkled in pale moonlight.

Jed rested in a dense stand of hemlocks. Meat supplies were cached in snow behind a quickly constructed camp. He gathered dead, lower branches from hemlocks to start a fire. Its warmth extended to an elevated place where he positioned his two snowshoes to form a springy cushion. From this resting place, with his back leaning against the trunk of a pine, he could look out over a warming fire and watch, in comfort, while the surrounding forest rustled with drifting flakes of snow. Undisturbed by other sounds the two travelers slept until dawn.

The first rays of sunlight glistened on a fresh layer of snow. The fire was rekindled to bring warmth back to camp. Sunshine formed a golden haze emblazing a web of snow-covered branches. In the snow, there were tracks of two wolves that had crossed an hollow in front of camp. Jed backtracked these prints to the back of his camp and discovered that two trout were missing. I could take more precautions when caching food, he thought. However, I'm not opposed to sharing what I have with forest animals. We share this life. The food should also be shared.

Following a breakfast consisting of roasted trout, Jed poured a cup of coffee then sat down to watch the forest. This is one of the places my grandfather showed me. His trap line went through here. Our community's traditional hunting area included much land around here and northward. I'm pleased to see that some things don't change. The land is still beautiful—just as in my memory. My grandfather gathered supplies and food from the woods; but did not collect more than was actually needed and not more than the forest could continue to supply. The need for sustainability is the idea the people of Clarksville have used to make their town successful again.

Jed's thoughts were interrupted by a distant, yet unmistakable buzzing sound of a chain saw. To Bow, he said, "We must go and see who is cutting our forest."

After walking through an area of hemlocks then hardwoods, the travelers came to an ice and snow banked stream. Gravel on the bed of this stream appeared through its clear, shallow water. While stopping with Bow to get a drink, Jed thought, maybe I should be cautious. I've already tangled with poachers. I'll keep my tracks concealed until I know what I'm approaching.

Jed returned to camp and prepared his equipment for traveling. As a precaution, he put a collar and rope on Bow then tied the rope to a tree at camp. Food and water were left for Bow who seemed to accept the situation. "If there's trouble ahead," said Jed to the resting dog, "you'll be safer here. I'll be back to get you. I also don't doubt your ability to chew through this rope if you get too distressed."

After leaving Bow at camp, Jed crisscrossed back over his trail to make pursuit difficult if there was to be any. When he reached the stream's bank, there was no sound in the forest except the gurgle of shallow water flowing over pebbles. A nuthatch walked down the side of an old stump by the creek's edge.

Jed stepped into the water. He hid his snowshoes and one pack under an eroded section of the creek's bank. Keeping his gun along with bow and arrows, he waded downstream. A strand of mist snaked above the moving water. Banks were eroded as well as sandy and topped by ice and snow. Along these banks, there were thick stands of cedars and hemlocks. Their branches often interlocked overhead, closing off much of the sky. Occasionally, small tributaries trickled into the main stream. All this water must come from springs with a little melted snow added in, reasoned Jed. The banks were cut deeply by well-used beaver paths, he noted. Deer tracks are also numerous.

The stream flowed into a valley of lofty white pines. This valley was flat at the bottom and divided by the creek. On each side, hills gradually undulated upward into a distant, blue haze. I've been here previously with my grandfather's trap line, recalled Jed. Why have I been away so long?

Sunlight had melted snow from rocks beside a falls. Jed climbed down these rocks then reentered the water. Mist hung in the valley above the stream and among walls of pines. Keeping to the shallow, gravel-bedded stream, Jed entered the valley.

Moving water kept ice from forming along most of the creek. A well-used raccoon path led to the water's edge. Along this trail, most snow had been melted away by sunlight. Snow, remaining on the trail, was packed and mud-covered. The path wound around the base of a particularly large, white pine's stump.

I know raccoons don't hibernate all the time, recalled Jed as he stepped out of the creek and followed the path to the stump. In its side, there was a large cavity extending upward. Even a person could climb in there, he thought. Probably generations of raccoons have lived here.

Without leaving prints on the trail, he returned to the water and continued walking downstream. Tracks of raccoons and beavers marked both banks. Trout occasionally darted through the shallow water.

Farther downstream, he heard the muffled sound of a chain saw being started then he saw two men kneeling beside a white pine log. They're having trouble starting the saw, Jed noted before he hastened onward. After observing more cuttings, he realized that the men were starting a logging road. A bulldozer was parked beside some trucks with markings stating, "Green Forest Logging Company". Two trailers housed a work crew. They're clearing a logging road along the bank of this creek, concluded Jed.

He continued walking in the water beside the road until he came to a region that had been clear-cut and thereby stripped of trees. Logging is a good business as long as the forest doesn't get cleared and destroyed, thought Jed. Then we have only destruction that is not good for loggers or anyone else. Any farmer or other harvester must protect the land in order to get another crop. If we log and replant some areas then the remaining wilderness can be left in tact for the enjoyment of others—including wildlife. A wilderness enjoyed by many forever is obviously better than a wilderness that is plundered by a few people for a quick, short term profit. Almost any business is okay as long as it doesn't wreck the environment. My family has used this land for generations; yet we have left no sign that we have been here. That doesn't mean we are no longer here or have stopped owning the land. I know this land has not been sold or leased for logging, particularly clear-cut logging. I must do something about these people who are breaking into my home. I don't want to harm anything or anyone. However, I believe in resisting an attack.

Jedediah turned back and walked upstream. While he passed the trailers and machinery, easterly—moving clouds blocked the sun increasing the grayness and chill of day. He approached the place where two men had been trying to start a chain saw. They were now sitting down with their backs resting against the trunk of a tall, white pine. One man was smoking a cigarette while the other used a file to sharpen the saw.

From his quiver, Jed withdrew an arrow and nocked it. I get weary watching people wreck the environment. There's the process of fracking that poisons water, along with the greedy exploitation of the tar sands and the oil leaks—and these practices too often have the support of the government. Before Clarksville was repaired, I watched people spray lawns with chemicals to kill weeds and insects. No one seemed concerned that these same chemicals also killed birds and animals as well as people. We have to work with what we have. In the case of these loggers, I'm going to send a protest from the forest to them.

Jed pulled the bow to full draw then released the arrow. It whizzed forward and slammed into the pine trunk to the side of the men and above them. They sprang to their feet. Looking wildly around, they just noticed what had hit the tree when a second, whistling shaft, followed by a third, slammed into the trunk. The loggers scrambled into underbrush and started running back toward their camp.

Jedediah waited until he could hear only the gurgling sounds of creek water splashing around his boots. As the sun slipped behind clouds along the horizon, he walked to a path the men used to get water from the creek. Hiding his prints on this trail, he proceeded to the trunk and retrieved his arrows, returning them to his quiver. He retraced his steps to the water.

Dusk was darkening the forest. Moving lights advanced from the direction of the trailers. Sounds of men's voices grew louder. Jed hastened upstream. Dusk darkened quickly into nighttime. Clouds covered much of the sky. Mist drifted above the creek. The presence of snow added visibility to the forest, outlining the creek's banks. By the sounds of approaching voices, the men are moving more quickly on their logging lane than I can travel along this creek, Jed warned himself. They could overtake me here, he realized as a cold claw of worry gripped him. If I move to the road, they will see my tracks. I hope I always have a way out of the corners I get myself into. This time, my only chance is that hollow trunk.

Hurrying onward, he came to the raccoons' path. He stepped out of the water, being careful to not leave definite prints on the trail. When his boot reached a root of the stump, he stretched around to the large cavity and climbed inside. He worked his way upward by pulling with his hands while stepping from one inner protrusion to another. He climbed until the opening became too narrow to advance any farther. With his feet secured on protrusions, he sat on another and rested his back against an inner wall of the remnants of this great, white pine. The raccoons will be farther up in the trunk, he told himself as sweat ran down his forehead. I don't think the men could see me even if they thought to shine a light in here. The interior of this trunk is rough and bent.

He waited and listened. Sounds of my breathing seem too loud, he thought. A piece of rotted wood dropped from above just before men's voices could be heard near the base of the trunk. I'm trapped, he exclaimed to himself. It's hot in here. Scuffling sounds came from the bottom of the trunk. Something moved above. I wonder if the raccoons are going to try to escape by climbing down my way, he thought. They must have an exit farther up.

A beam of light flashed into the lower cavity. This light appeared briefly then darkness returned with an increased intensity. Sounds of boots crunching on snow moved past the stump. While walking away, one man said, "Bill and Len didn't see anyone. We haven't heard anything or found any tracks. There are no arrows in the tree either. Bill said they didn't look like the kind that can be purchased in a store. The arrows looked like the old kind. Maybe we have stirred up something we should leave alone. Let's go back." Sounds of boots crunching on snow faded until only an eerie silence remained.

Dim light entered the interior of the tree from the opening at its base. Some light also entered from above and outlined rough features of the narrow upper passageway. The light above my head was not always there, he said to himself. Raccoons must be stirring. Occasionally, they move and block the light.

While Jed watched, raccoons left the hollow trunk through an upper opening. Some scratching sounds could be heard on the tree's outer surface as the animals climbed down to the ground. After their departure, more light entered the cavity through the upper opening. Flakes of snow also drifted inside. While flakes added a fresh coating of snow to the forest, Jedediah slept inside the aged white pine.

He was startled when he woke up because he felt trapped until he remembered where he was. A bright patch of moonlight marked a section of the upper hollow. An owl hooted loudly.

Sleeping again, he awoke to the fluctuating roar of a chain saw's motor. No light came from the upper channel. Only from below did dim light outline the rough interior. Slowly, he climbed down to the base of the tree. He was chilled by a cold grip of worry. I can't leave Bow at camp much longer, he reasoned. I have accomplished nothing here. I'll try again then leave quickly.

With carefully placed steps, he worked his way back to the creek. He entered its water and walked downstream. A fresh coating of snow topped banks and trees. The white forest shone brightly and coldly against a gray sky.

He walked downstream and saw the same two men fixing, or sharpening, the saw. That first message I sent was not heard, reflected Jed. I have accomplished nothing. I'll send another message then leave. As he had aimed the bow the day before, Jed now sighted his rifle above the men. Bullets screamed over the men in a burst of gunfire that exploded then echoed through the forest. The loggers dove for cover in dense underbrush. They kept running in the direction of the trailers.

Jed rushed downstream and went past the trailers. He watched six men assemble in front of these structures. The loggers carried rifles and hastened up the laneway toward the area where the shooting had taken place.

I could be making the biggest mistake I've made so far on my journey, he warned himself. I have to stand for something. Maybe, though, this isn't the way to do it. I know what I stand for. I just have to do something. He fired a few shots, sending bullets whooshing over the camp. The loggers returned and took cover behind machinery as well as trees. Shots were sent in Jed's direction, covering a wide area. Bullets whizzed overhead.

Jed kept below the creek's bank, occasionally sending shots over the camp and shooting each time from a different location. An exchange of rifle fire settled into an intermittent volley of shots whistling above the camp and creek.

Jed let the loggers do most of the shooting. They won't know how many people they are confronting, he reasoned. I've been protected by the creek's bank. I've kept my shots high and left no tracks. I should stay upstream so I don't get cut off from my back trail.

Just as the thought of being cut off occurred to him again, he saw three men run from behind the trailers and move in a direction that could block the upstream route. Jed reacted quickly and hastened along the creek, keeping as much as possible beneath the crest of the bank. Knowing the men could use the bush road to get ahead of him, he was relieved to see the hollow tree looming up over the creek bank. With the same careful steps, he went inside and climbed to his resting place.

Jedediah listened to the sounds of his own breathing and the crunch of footsteps in snow as loggers walked close to the tree. He also heard a man say, "There are no tracks—never have been any."

Jed slept inside the tree until he was awakened in the afternoon by a whirring buzz of a chainsaw's motor. During my journey, he reflected, I seem to have changed from a procrastinating thinker to a thinker who takes action. I must make sure Bow is safe and take some time to think about what I'm doing here. He climbed down to the opening. Cautiously, he left the tree and was careful to leave no boot prints on the path to the stream.

A cold grayness settled into the afternoon while he walked upstream. Clouds, mottling the sky, covered much light from the sun. The climb at the falls seemed to be more difficult than the descent. Upon reaching a poplar that had fallen across the creek, he pulled himself onto the trunk and left no tracks as he walked away from the water. When prints in the snow could no longer be avoided, he backtracked twice before approaching his camp.

Jed first heard barking then Bow came into view. He was running with a rope trailing behind him. At close range, the dog jumped at Jed, knocking him over and barking at his face. After a wrestled greeting, Jed said, "I was away too long."

Jed returned to the eroded bank and uncovered his equipment. Back at camp, he put two frozen trout on a log near the route the wolves had taken previously. In addition to some fillets, two large trout remained. The two trout were skewered over coals. Water was boiled for tea. This meal restored a feeling of contentment to Jed. Bow rested while Jed enjoyed a cup of tea. As evening shadows darkened, the flame brightened the camp. In the distance, a wolf howled. Both dog and man slept until the gray light of dawn brightened the forest.

Bow circled the camp in a burst of excitement when Jed, having packed his equipment, was ready to travel again. First rays of sunlight emblazoned edges of clouds with red light while adding similar tints to a rugged, winter landscape. The northern route stretched into higher country amid hills. At the edge of a swamp, Jed noted that the snow's surface had been colored brown by the scattering of bulrush seeds. After entering this swamp, Jed kindled a fire by starting with an handful of birch bark topped by dry pieces of cedar.

A light film of oil was first heated in a frying pan before Jed added the flour-covered trout fillets. They sizzled in the hot oil. Bow waited patiently for an even share of golden fillets served on birch bark. To his own serving, Jed added a sprinkling of malt vinegar followed by a little pepper. Bow drank water from a spring. Jed prepared tea and enjoyed the satisfaction of having made a comfortable camp amid wild surroundings.

The next day brought a new start to the journey north. They crossed moose tracks occasionally mixed with wolf prints. The terrain became increasingly rocky. Snow covered hills were forested with white and black spruce as well as white and jack pines interspersed with poplars and birches. Bow chased snowshoe rabbits that seemed to vanish beyond an initial spray of snow kicked up as they ran. At the end of one particularly long sweep of snow-covered rocks, Jed and Bow came to the edge of a cliff. Overhead, an hawk circled slowly in a light blue sky. Below the cliff, treetops swept gradually down into a valley cut by a dark blue ribbon of a river. Along its banks, there was a village of wigwams. From most of these lodges, smoke rose into calm air then hung in an horizontal, blue strand above the wide pocket of land. A traditional village, exclaimed Jed to himself. I didn't know this was here. To the left of Jed, an hawk plunged toward the earth. Jed watched the bird move beyond treetops. While watching for the bird, he saw a person approaching. The man wore boots, jeans and a leather jacket topped by a fur robe and hat. His black eyes were alert as if the whole forest was of particular interest to him. The rifle he carried was held like just another piece of necessary equipment and not a threat to Jed. The stranger said, "I've been watching you. What do you want?"

"I didn't know you were here," admitted Jed. "I'm Jed Speaker. Who are you?"

"I'm Charlie Fisher," he answered. "This valley is a traditional village. It's one of the ways we are keeping our customs. Most of us are from Heron Cove. I've heard about Tom Speaker having a grandson who had gone south to Clarksville. Tom is here in our camp. He's with some other elders and the minister from Heron Cove. You would be his grandson?"

"That's me alright," answered Jed. "I've been away a long time. But I'm on my way back now."

"Tom has a wigwam at the east end of the valley," explained

Charlie, pointing toward the end of the camp. Moving his hand in the opposite direction, he added, "Walk to the west along the cliff and you will find a path. We have police, like me to watch this place. You never know who's around. Trouble can come from inside the village or outside. Loggers have moved in from the south. They have no right to be cutting trees there. We've had trouble with them. They don't realize that the law is on our side. We've taken our struggle to the courts and it's there that we are defending our land against not only logging companies but also mining and oil companies. We support development that does not wreck the environment. We don't want this wilderness to become a clear-cut garbage dump like the tar sands in northern Alberta where the environment comes second after greed and ruthless, short-term profit."

"Has the village been here long?" asked Jed.

"The village site has been a camping place for as long as anyone can remember," answered Charlie. "This valley, in earlier times, was used during the winter. Now we keep the camp occupied all the time. Old pieces of pottery bowls along with flint arrowheads and spear points have been found. I discovered a copper knife in a sandy riverbank. Someone found a piece of conch shell in a sand bank. Such shells would have come in from southern trade routes. Trade routes extended in all directions. There are also clay pipes in camp that have come from the time of the European fur trade. Most people who stay here now are visitors. They come for cultural and spiritual renewal. Tom Speaker said that in a vision he saw his grandson coming home. Likely, he's expecting you."

"And I thought no one knew what I was doing," exclaimed Jed. "I'll find the wigwam—and thanks for your help."

"Pleased to meet you," replied Charlie. "Can't say it was a surprise." He turned away and started walking back the way he had come.

Jed followed the cliff's edge. Bow had been checking tracks along the back trail and met Jed at the path. It followed a boulder-strewn gully dropping to the valley floor. This trail continued between walls of black spruce then came to a clearing occupied by a scattering of wigwams.

Two women seemed to be enjoying the task of cutting firewood with a chain saw. A girl played with four pups beside the river. Silently, she watched Jed. The women stopped sawing and noted Jed as he put a collar along with a rope leash on Bow. After finishing his work, he asked the women, "Do you know where Tom Speaker is?"

"I haven't seen him today," answered the older woman who had been operating the saw. She wore a leather jacket over a dress. Moccasins and leggings protected her feet and legs. A kerchief bordered her lined and friendly face.

The younger woman wore jeans and boots along with a fringed, leather jacket. Her long, black hair shone in sunlight. "His wigwam is the last one next to the river," she said, pointing toward the end of the village.

"Thank you," replied Jed. He continued walking with Bow secured to the lead. Three young women walked past. Briefly, as if they were marching, their legs moved in unison. From some lodges, voices could be heard along with occasional laughter and some singing. Drums and rattles accompanied the singing. Wood smoke drifted in the air together with a tantalizing mix of cooking aromas.

Maybe I should not have come here, thought Jed as he approached the last wigwam. I hardly know my grandfather. At the lodge's open doorway, Jed looked inside. He didn't recognize the gray-haired and bearded man with bluish eyes peering alertly through metal-rimmed glasses. He wore a black coat over a woolen sweater. They fit loosely as did the jeans and moccasins. "Are you Jedediah Speaker?" asked the man.

"Yes," answered Jed. He was about to walk away and try another lodge or leave the camp when the man said, "Come in. We've been expecting you. I'm Cal Tomkins, minister at Heron Cove."

"Do I tie the dog outside?" asked Jed.

"If he's friendly, bring him in," said the man.

Jed and Bow entered the comfortable lodge. It was scented with fragrances from balsam boughs and wood smoke. A small fire sent a trail of smoke up between poles at the top. Jed sat across the fire from the man. Bow stretched out beside Jed.

Cal's eyes portrayed calmness of a person who had seen much of life and could look at any worrisome situation with a calming wider view. "Have some hot, roasted, moose meat and wild rice," he said while preparing a plateful of fragrant meat topped by long, dark grains of rice. The food was accepted gratefully. Jed had not realized how hungry he was until he saw the meat and rice. He received a second dish of cooked meat as Cal said, "Here's some food for your partner. I have water for him. Tea will soon be ready."

"Thank you for this treat," exclaimed Jed enthusiastically. My dog's name is Bow. He's appreciating his food also."

Following a fine meal, Bow rested while Jed and Cal enjoyed some tea. "You are my grandfather's friend and also a minister," said Jed.

"Yes," replied Cal. "We work together. Tom said you were traveling north. He hoped he would see you here. Today he went north to the ridge. To get there you follow the path eastward to a bridge across the river. Keep to the trail and it will take you up the side of the valley to one of Tom's favorite places. You should visit him first then please come back here."

"You two seem well organized," noted Jed.

"We try to be," replied Cal. "Some people think we don't do anything. Actually we're busy all the time. When life runs smoothly, the work that contributes to the smoothness seems easy or not there at all. And, you know, there's no such thing as coincidence."

"Thank you for your hospitality," said Jed. "Should I go and see my grandfather now?"

"When you are ready," answered Cal. "Please come back here afterward."

"Thank you again," said Jed as he stood up and started carrying his equipment outside. Bow got up at the same time and was the first to leave the lodge. Before stepping away, Jed called back to Cal, "See you soon."

Jed kept Bow on the rope leash. They walked into sunlight brightening the valley. The trail led to a place where the river was covered with ice. I guess this is the bridge, said Jed to himself while he followed a path of footprints to the opposite bank then entered a forest of black spruce. The trail avoided a rock wall by veering to the side and climbing a gully to a summit of flat, snow-covered rocks bordering the valley. At the cliff's edge, overlooking treetops sloping into a blue haze, there was a campfire. Sitting beside this flame was a man whose features Jed vaguely remembered. The old man had prominent cheekbones in a lined face bordered by gray hair protruding from the edges of a muskrat fur hat. Although the face seemed older, the eyes had not changed. They held the same fire as Jed remembered from the times Tom Speaker had come to visit and tell his wondrous legends.

The eyes flashed with recognition when Jed approached with Bow. They both sat beside the fire. "Good to see you again," said Tom while pouring Jed a cup of coffee. "Been a long time. They all seem to prefer tea around here. I prefer coffee. Do you mind the stuff yourself?"

"I prefer it too," answered Jed receiving a cup of particularly black coffee. "Thank you."

"I've been watching an eagle," said Tom before he lit his pipe sending a strand of smoke curling toward the blue valley. "The ceremony, I enjoy. I don't inhale the stuff and only use the pipe when I'm up here. I don't smoke—so to speak. I've learned how harmful the smoke is. Seems everything's harmful nowadays—especially food. Finding healthy food is as difficult as finding an environmentally friendly oil executive. Do you smoke?"

"No," answered Jed. "I hear you knew I was on my way home."

"Yes," said the old man. As usual he was selective in his explanations.

"I walked to your lodge and met Cal Tomkins," said Jed. "He told me you were here."

When Tom pointed skyward, Jed saw the eagle and both men watched as the majestic bird circled overhead then Tom asked, "You here to stay this time?"

The old man's voice had a raspy ring to it that seemed to be as much filled with mystery and wisdom as the legends he told to Jed when he was a boy. "Yes," answered Jed. "I seem to have been away a long time—too long. While away, I've learned what I do not want to do and there's some value to that."

"We are always learning something," responded Tom. "I have wanted to make sure you got started in the right direction. You should keep your traditions and get to know the Creator and thereby get to know your own path that you will follow to experience and develop before you cross over again to the spirit world. We are individualized parts of the Creator. Beyond that awareness, we learn and develop, learning how each person is unique and also part of all the rest of creation. Birds, fish and animals are pure spirits, just as they are. They aren't refining themselves and developing like the human beings. We are all, though, connected and part of creation. Unlike creatures, some people willfully turn away from the Creator—and that's the only devil there is. That's my advice to you. The rest of it is just in the details."

"Thank you," replied Jed. "I'll take time to think about what you've said."

"Probably a life time," said the old man, smiling.

After drinking some coffee, Jed asked, "How are my parents?"

"They are well," he answered. "Your old girlfriend, Carol, has been well also. She is a nurse now. She has not forgotten you and has asked me about you."

"I have not forgotten her either, grandfather," replied Jed. "How have you been?"

"Been well most o' the time—except when I'm sick," he said, smiling brightly. I don't get older. I just get tired more easily. I haven't been trapping as in the earlier days when you helped me south of here. Now there are less animals and loggers have invaded my land."

"I saw the loggers when Bow and I walked here," said Jed. "My dog's name is Bow. He's good company."

"I've always liked dogs," observed Tom. "I know a lot of legends about dogs. You never know when you are going to notice a coincidence that is really divine intervention or more noticeable presence of the Creator such as a person sent at the right time. I once met an old man up here. There are legends about such happenings. The meeting did not surprise me. But, of course, some people say I'm just an old man." The smile returned.

Sometimes I'm not sure when Tom is joking or when he is serious, thought Jed. "What was the man's name?"

"The name doesn't matter, really," answered Tom who seemed surprised by such an unnecessary question. "He said his name was Caleb Pine."

"He's the one who saved Clarksville from destruction caused by pollution," said Jed. "Seeing the revival of the town got me to realize I needed the same change on a personal level. That realization started this journey home."

"Welcome back," exclaimed Tom.

"Your legends are usually about the old days," said Jed.

"Nothing has changed though," replied Tom. "The important things don't change. I hope I will see you more often now that you are home again. I don't know how long Cal Tomkins will be staying today. Maybe you should talk to him before he leaves."

"I didn't know I was coming here to learn," said Jed. "I thought I was just going home."

"You're doing what we're all doing," said Tom. "Now that you are home, don't be a stranger—to your home and to your own life."

"Stranger to my own life," repeated Jed. "That's exactly what I was doing."

"When you see familiar guideposts, you know you are on the right course," stated Tom.

"I'll go and see him if you want me to," said Jed as he stood up and placed the empty cup on a stone beside the fire. "Thanks for the coffee. Good to see you again, grandfather. You haven't changed."

"I hope that's good," exclaimed Tom. "Your parents will be happy to see you—and Carol also will likely be expecting you."

"See you soon," said Jed before he and Bow started walking back to the trail. It returned them to the riverside lodge. From its top, a trail of smoke climbed the sky before bending with similar strands to form a slight gray film over the camp. Above the valley, beyond the smoke, an eagle screamed as Jed and Bow entered the wigwam.

"Congratulations," exclaimed Cal while Jed and Bow sat down beside the fire. "Tom said he wouldn't send you here unless he thought you were ready. Advice that is not requested is too often just discarded. Have some tea. There is water in the dish for Bow."

Jed poured tea for himself and refilled Cal's cup then they all rested beside the flame as it flickered above coals and two recently added pieces of wood.

When Cal was ready to talk, he looked into the fire and said, "Jed, like everyone else, you have made mistakes."

Surprised by this direction to the conversation, Jed was concerned about what might be coming next and was relieved when Cal continued to say, "Those mistakes have started you on a journey here. The real journey, though takes place inside yourself to discover not only your own uniqueness, and accordingly, your special purpose, or purposes. We each have are own individual path to follow for experience and development in a refining process. Tom and I follow two paths to the Creator and our role is to point them out to you. We are your guideposts and that's why you are not surprised to see us and we were expecting you. One path leading to God can be found in the teachings of Christ and another is presented through the traditional Indian beliefs. Paths leading to the same place are, to themselves, similar. Tom and I follow both trails and we recommend both to you. That's why the three of us are here."

Jed felt a spirit of peacefulness in the lodge he had not noticed previously. He also knew he would be aware of this presence for the future. He had come home. Now the rest of the progress involved adding the details. The words he had heard on the cliff and in this lodge, he, to some extent had heard before and they had not resonated with him. Today the words had become emblazoned with light, leaving him at peace with his part in being connected to the rest of life. With such an elevating connection, he knew he was home and the feeling would remain regardless of where he traveled. He could not get away, and did not want to get away, from his true self and the joy of being with the Creator as an individualized part of one life.

Listening again to Cal, Jed realized that the man had been speaking and was saying, "The only hell occurs when a person, deliberately, by his or her own choice, turns away from the Creator."

"I certainly thank you and my grandfather for these messages you have given me today," said Jed. "Your messages leave me with a question for you. On my journey here, I had time to consider information I had been told in the past. I learned to respect the environment and the importance of taking care of it. I ran into some loggers who are cutting a road and starting to clear-cut Tom's land south of here. These loggers are like the mining and oil companies, and the governments that support them, that don't believe in having both employment and an healthy environment. Motivated only by short—term profit they trick people into believing they can't have both an healthy environment and jobs. These liars say it's okay, maybe even necessary, to wreck the environment to support employment and prosperity. I tried to stop these loggers and I failed."

"How did you try to stop them?" asked Cal.

"I used arrows first then bullets," he replied. "I wasn't trying to hurt anyone and didn't. I just wanted them to stop wrecking the forest. Logging can be a good business. These people, though, were destroying the forest."

"Why did you not ask for the Creator's help?" asked Cal.

"I didn't think of that then," he said.

"Tom and I work this way all the time," replied Cal.

"I have enjoyed my visit to this village," said Jed. "I will never forget it."

"You are welcome to stay," said Cal. "I'll be leaving soon. You could stay here with Tom or there are other lodges."

"Thanks for the invitation," he said. "I'll be back again. I haven't finished my journey yet." Standing up, he added, "I've enjoyed meeting you and seeing my grandfather."

Handing Jed a parcel, Cal said, "Some moose meat for your journey."

"Thank you again," he replied, putting the parcel in a pack. "See you soon," he called back before following Bow out through the doorway.

They left the village and climbed the gully to reach the cliff then they turned to the south. The snowshoes were supported by an hard crust of snow beneath a fresher layer on the surface. Occasionally a spray of snow would appear and Bow would chase a fleeing snowshoe rabbit. On high points of land, trees were often stunted and twisted. Valleys had dense stands of black spruce or white pines.

When Jed came close to the valley where the loggers were cutting a road, he built a camp on an hill topped by white pines overlooking the ice and snow covered surface of a small lake. The metal stove kept the tent warm while a fire outside was maintained for most of the cooking. Cedar boughs on the tent's floor added a fresh, resinous fragrance.

Jed roasted some moose meat over the outside fire. This tender food was shared with Bow. Afterward, tea was steeped. Jed savored some tea then decided to visit the logging road.

With Bow, Jed headed south and came to a stream that flowed from the lake and entered the valley in a cascading ribbon of a waterfall that plunged down a face of a cliff. Walking west, Jed reached an high point of land topped by white pines. This stream enters the creek I followed in order to confront the loggers, thought Jed as he scanned the valley. From here, I can see the main creek and also the logging camp. I'll make a secondary camp here.

Concealed by surrounding rocks and pine trunks, Jed built a campfire using dry hardwood providing a smokeless flame. He steeped more tea then leaned his back against a pine trunk and rested while enjoying the warming tea. Everything I've done in my life so far has likely been a mistake except for that visit to the traditional village, he mused. After visiting the village, I continue to be alone although never completely so again. I might be about to get into more trouble. However, it would be a mistake to do nothing while people are killing the land. Anything that harms the land also harms people. More than one group draws life from the forest. My ancestors have legal ownership of the land because they were here first—the first occupants. They received it from the originator, the Creator who also continues to be present. People need money. But only those who have more greed than sense would kill the farm to get the first crop. As Cal Tomkins said the power of the forest could be tapped for its own defense. Jed offered tobacco to the fire, sending the smoke to accompany his first prayer that the loggers would stop logging.

Jed waited with Bow at the secondary camp although he noticed no change in the forest or logging camp in the valley. He perked more coffee then sat down to enjoy it. A light seemed to flash in the forest making each part appear much brighter and clearer as if transforming to a sudden transparency allowing him a glimpse of the spirit side of infinity and beauty to which he was a part, connected to each other part. He would never, he knew, be the same again, having seen the more important, spiritual side of life, the only part that mattered. The vision over, he continued to notice that his surroundings were more vibrantly filled with life. I've become more aware of things around me, reflected Jed while he watched the landscape with renewed fascination. The logging, however, continued in the valley. The increased light seemed to fade again although not entirely.

Jedediah and Bow walked to the lake beside the main camp. After chopping an hole through the ice, Jed tried jigging and almost immediately a fish struck the lure. In a spray of water, a trout was pulled from the hole. Along the fish's silver sides, rainbow hues flashed in sunlight. More fish were caught as flakes of snow drifted from clouds etched in golden outlines by the sun's rays. Moving snowflakes partially obscured the shoreline patterned with rugged black spruce and white pines.

Vaguely, through swirling snow, Jed saw two wolves approaching. They were large black timber, or gray, wolves. They came close enough for Jed to notice a patch of white fur on one wolf's front leg. Farther back, behind the first two animals, there were four other wolves.

Jed secured Bow by attaching his collar and leash then tying the leash to a pack. Next Jed put a stick through the gills of six trout. He carried these fish toward the wolves. As Jed approached, all the wolves stood motionlessly on the lake's frozen surface where snow swirled in wisps that appeared and vanished fitfully. The same breeze stirred tufts of fur on the wolves. The wolf with a partially white leg remained in front of the others although they were all in sight when fish were scattered on the snow. Jed returned to his fishing place to watch with Bow.

The wolf with a white leg came forward, picked up a trout then started carrying it to the far shore. Hesitantly at first then swiftly each of the remaining animals came onward and took a fish before running beyond an obscuring veil of snow. When shadows of evening started appearing on the lake's surface, Jed stopped fishing. With a catch of twelve more trout, he walked with Bow to the main camp.

A red swath of moonlight crossed the lake. This reddish sheen turned to silver as the moon rose above a snow covered landscape. Jed roasted trout for himself and Bow. Afterward, more wood was added to the fire. A tall flame flickered into the cold night. Jed prepared green tea and sipped it contentedly while watching the moonlit wilderness. A fox carrying a ruffed grouse walked past the camp. The departing hunter blended with and quickly vanished amid many other forms in a darkening forest beyond the fire. In the direction the fox had walked, an owl hooted.

While sipping tea and watching the fire, Jed thought, maybe I'm reverting back to my old habit of procrastinating more than acting and not doing anything until I've run out of choices. Possibly I shouldn't be in the wilderness—almost alone—when any injury could be serious. People see the forest's beauty and are drawn to it. Once they get into the woods, they feel uncomfortable with its wildness so they start removing this wild beauty until they feel at home. The forest is like air. It's essential for everyone—not just one group. Each aspect of the wilderness is unique and connected to every other part. Logging can be a good business if the process does not destroy the forest such as occurs with clear-cutting. This local company would remove the forest. Logs preferably should come from a tree farm rather than a wilderness. In a tree farm, an area would be clear—cut then replanted. Here the wilderness is destroyed by being clear-cut then left in ruins of not much further value to anyone including loggers.

After watching the campfire and surrounding, moonlit landscape, Jed, then Bow, slept until the rising sun sent shafts of light across the snow. Roasted moose meat provided a fine breakfast for both man and dog before coffee was perked.

In sunlight, Jed and Bow walked to the lookout point to check on the logging camp below. There were the same trailers, trucks and bulldozers. Work seemed to be continuing as usual.

While Jed and Bow returned to the main camp, croaking calls of ravens rang from the sky overhead. Large wings brushed against warming air while the birds chattered and seemed to be watching the dog and man below. A flock of pigeons flew above treetops. Each bird turned and maneuvered in unison. Sunlight flashed on feathers emblazing the flock against the sky.

With a booming sound of rapidly moving wings, some ruffed grouse took to flight near camp. Shortly afterward, the black wolf with a white patch on one leg appeared carrying a grouse. The animal crossed an open area then entered a dense stand of black spruce. "Some people consider the wilderness to be a dangerous wasteland," said Jed to Bow. "I don't find this to be a threatening place—as long as it's treated with respect."

A meal was prepared consisting of fried, trout fillets, cooked until golden and crisp on the outside while being moist and tender on the inside. Jed and Bow shared the fillets before coffee was perked. Bow slept while Jed poured a cup of coffee and sat down beside the fire. This day is getting milder, he observed while watching the forest. Water had started to trickle into low areas. One or two thaws usually occur each year, he recalled. Although the woods can be a rough place, a person with the right equipment and knowledge can enjoy the place to an increasing extent until its beauty seeps by degrees, and also flashes, into the soul, leaving an indelible spiritual awareness that is calming and restful. Instead of allowing city influences to destroy the wilderness, we should be trying to improve the environment of the city. We should fix something that needs fixing rather than wreck something that is working well and has been working well for countless centuries.

When the weather became mild, melting snow dropped from trees. A Canada jay came for food and was followed by a flock of chickadees. Although the evening sky was clear, wind-blown clouds approached from the west. They gradually covered the sky while wind turned flames of the fire toward the east. By nighttime, rain was pelting the snow-covered landscape. Jed secured the tarpaulin over his camp.

After the wind calmed to become a slight breeze, Jed again

enjoyed the comforts of his campsite. The tarpaulin provided extra shelter from the rain. The fire in front of this tarp helped to provide dryness and warmth. Firelight became the only touch of light amid the murky forest. Such illumination briefly outlined the form of an Arctic owl as it glided over the camp.

Lightning caught Jed by surprise and made Bow nervous. I haven't seen lightning many times during the winter, recalled Jed before a crash of thunder rocked the forest and further agitated Bow. The night brightened again and then often as patterns of lightning followed by explosions of thunder gripped the forest. This storm increased in force with a raging wind that snapped trees.

Jed took down his tent in addition to the tarpaulin. He found shelter for himself and Bow in a rocky hollow. Protected by these rocks, Jed tried to calm Bow as the storm emblazoned the sky with lightning amid a din of thunder and ravaging wind. The wild night gathered itself into a crescendo of a powerful explosion of wind blasting out of the sky and rocking the forest. Jed knew he and Bow had missed the worst of the blast by being sheltered in the cave-like hollow.

When the storm started to diminish, Jed added the tarpaulin to the rocky hollow then built a small fire. Both man and dog slept until the first light of dawn. The morning was colder than usual. First rays of sunlight glimmered across a frozen forest. A breeze caused creaking and rustling sounds when branches stirred under a coating of ice.

With Bow, Jed walked to the elevated area overlooking the logging camp. He used binoculars to check the valley. Men stood around a fire in the center of what had been their camp. Tangled pieces of trailers littered the area. Trees had toppled onto machinery. While Jed watched, the men left their fire and started walking out of the valley. Hopefully, reflected Jed, after all the difficulties the men have faced here, they will not return. I used prayer to have them stop logging and they stopped. However, they have left three dogs behind to wander as strays. There will be abandoned food in camp to last the dogs for a while.

Jed stayed at the lookout site for a few days to observe the ruined camp. The dogs had left soon after the departure of the men. Neither men nor dogs returned. The forest has reclaimed its valley, mused Jed as he walked back toward the lake near his main camp. On the snow-covered ice of this lake, the three loggers' dogs watched a bull moose. Surrounding snow was trampled and packed with prints. The combatants are tired, noted Jed. By marks in the snow, this standoff has been going on for a long time.

Catching the dogs off guard, the moose bolted forward and trampled one attacker then another. The third dog sprang for the bull's hind leg and was met by a kick that tumbled the dog into an unmoving heap. The bull stood over the three adversaries. Snow in the area was patched with blood. The bull's back legs were torn. The large animal seemed to droop as the fury of battle and blood drained away. After trying to walk once, the animal stopped. Its hind legs are ruined, reasoned Jed. I don't like to knock down a champion. However, he can't do anything more the way he is. Jed fired the rifle once and the moose slumped to the ice.

Flakes of snow whipped across the ice while Jed pulled the dogs to shore before returning to salvage the extensive supply of moose meat along with the hide. Meat was placed on the hide and pulled to camp.

Sheltered by the tarpaulin, Jed built a large fire to roast meat and steep some tea. Although flakes of snow swirled across the landscape, there was shelter, warmth and food at the camp. With a cup of tea, Jed rested and watched the moving veil of snow.

By evening, the sky had cleared. A flame jumped from a deep bed of coals and flickered light across the camp. At the edge of this warmth, Jed slept soundly following a meal of roasted moose meat. He also slept in a sitting position touched by warmth from the coals. An owl's hooting awakened him at night in time to add more wood to the coals. He slept again until the grayness of dawn broke along the eastern horizon. While dawn's light crept across the land, the usual solitude was broken by a distant, buzzing sound of a chain saw. "That can't be happening again," exclaimed Jed to Bow. Jed felt a rush of anger mixed with frustration. "My plans have often been thwarted," he said. "I've seldom though, in the past, experienced a pulse of anger that I feel now. We should have breakfast first then we'll see what's happening."

A fine breakfast was cooked over a wide bed of coals. Coffee was perked and enjoyed while all equipment was readied for traveling.

Walking was exhilarating on a fresh layer of snow that blanketed the forest's floor. Clumps of snow streamed down from trees when gusts of wind moved branches. Squirrels running along branches also dislodged curtains of snow. With a booming of wings, ruffed grouse took to flight from a stand of hemlocks. Beneath lofty, white pine boughs silhouetted against a clear sky, Jed and Bow reached the height of land overlooking the valley. It glistened in shades of silver and gold where rays of sunlight touched the forest. A raven croaked as it flew overhead. Afterward, the solitude was broken only by a persistent buzz of a chain saw's motor.

Using binoculars, Jed noticed that a new work crew was rebuilding the logging camp. With more equipment and trailers, these men were reopening the road, clearing away the wreckage and establishing a new camp. Rather than help the forest, I seem to have made a larger enemy, thought Jed in puzzlement and disappointment. I tried to accomplish something and seem to have achieved nothing. Like most issues, this struggle appears to come down to a position of strength. People who would destroy the land such as logging, mining or oil companies, too often seem to have more strength than protectors of the environment.

Jed built a fire, perked coffee then sat down to sip this warming drink and consider his next steps. The Creator has all the real power and life, reflected Jed. However, we have to know how to become part of these strengths. First I assumed others would see the varied benefits of the wilderness and recognize how they help everyone. These particular loggers do not recognize the value of both an healthy environment and long term employment. They just see quick profits regardless of a greater cost in terms of environmental destruction. The loggers limit their view. I have made the same mistake because I've limited my view and my effort. I just tried to have them stop their destructive work and I seem to have failed. Jed poured more coffee and said to himself, I have been dealing with unnecessary limits. I will now ask the Creator for the complete takeover of this logging company. Producing logs or furs and other products from the land are good business activities if the environment is also protected as a priority and not as an afterthought. The company here is destroying the land and this is the problem I'll work at solving. There are, I'm sure, larger problems to be tackled in other places by people who have their own particular qualifications for undertaking such tasks. I, like others, can only do my best in my present location.

Jedediah dropped tobacco into the fire. With rising smoke, he sent his larger prayer. Smoke ascended while an eagle soared above the valley. I will now, Jed told himself, return to the place where I built a sweat lodge.

Jed and Bow returned to their main camp beside the lake. As much food as could be carried was packed for a journey that was started immediately. Jed started walking toward the south.

Bow was a tireless traveler. Once he knew the direction Jed was taking, Bow ranged throughout the area, checking scents and following tracks. Because he enjoyed discovering new country, Jed veered westward and came to an area of particularly high, rocky hills. Undulating slopes topped by white pines overlooked valleys of black spruce. In a clear, blue sky, ravens frolicked. Sunlight flashed on their broad wings.

As Jed picked his way through the forest, he noticed a crow at the top of a spruce. The bird cawed repeatedly when Jed approached the tree. There is trouble here of some kind, thought

Jed. Maybe the crow has become tied in some way to the tree. Like people, birds and animals have their share of mishaps in the wilderness.

Jed climbed the tree slowly, but steadily, picking his way from branch to branch until he could reach the bird. Its feet had become tangled in fish line that had become tied to a branch. The bird watched quietly while line was cut and the feet freed. Cawing repeatedly, the released bird flew to a more distant tree.

After climbing down from the spruce, Jed decided to make a camp. Using dry, spruce wood, a fire was soon started. On a distant stump, Jed placed some meat to help the crow. That bird was weak from lack of food, thought Jed while he walked back to camp. A comfortable sitting place beside the fire offered him a wide view of the spruce forest including the stump with its meat. The stump's first visitors were chickadees then the crow arrived. It fed on the meat before flying into a graying dusk. Shadows gathered upon the snow and gradually darkened with the approach of night. Wolves howled in the distance. Afterward, silence was unbroken. Jed and Bow slept until dawn when resonant croaking calls of ravens announced the start of a new day.

First rays of sunlight found Jed and Bow continuing their journey southward. They traveled slowly, camping beside a snow etched swamp then beside a cliff. Further campsites in sheltered valleys of black spruce eventually led to an high falls where water cascaded over curtains of ice. Beside this falls, on sheer, rock walls, Jed saw pale, reddish pictographs.

Ice did not cover much of the stream beyond the falls because the current was too swift. When filling his coffee pot, Jed noticed a flint arrowhead on the gravelly bed of the creek. Picking up the finely shaped piece of flint, he thought, what a story this stone could tell. It has been made and used, in early times, by an Ojibway hunter.

While Jed noted the finely chipped shape of the arrowhead, feathers drifted through the air in front of him. Following the path of these feathers back the way they had come, he saw an hawk on a branch. The large bird was busy tearing feathers as well as meat from a freshly killed pigeon.

Coffee was prepared in an hastily constructed campsite. Moose meat was also roasted. Overhead, a scattering of clouds turned crimson in rays from the setting sun. The fire flickered more brightly after the red circle of the sun dipped below a rugged horizon.

After breakfast next morning, Jed and Bow reached a large swamp. It was a maze of old willows. Here the solitude was broken only by the sound of an hairy woodpecker tapping intermittently on a stump. Ice and snow covered ponds were patterned by tracks of squirrels, rabbits, deer, foxes and wolves. Each line of prints was a new source of interest for Bow who had become a tireless wanderer. The swamp's more secluded places revealed prints of ruffed grouse, turkeys and mice. One mouse track ended at an hollow in the snow where an owl had swooped to a kill. Wolf tracks led to a creek. Its water was not covered by ice because of the current. Ice bordered the banks while water murmured swiftly and clearly over a bed of sand, gravel and pebbles. Upstream, Bow led Jed to his old sweat lodge. Fox tracks circled the lodge site and crisscrossed the area. I'm not sure why I came back here, reflected Jed. I remember, however, the special peacefulness present in this place as though it is one of the thin places on earth where the other side, the spirit world, is more noticeable.

Jed removed snow that had drifted inside the lodge. Afterward, he built a fire and put up his tent. The surrounding forest continued to emit an almost eerie calmness. Cedars blocked most breezes. Along the creek's banks, sunlight had melted sections of snow and ice, revealing sand. The gravel in this creek bed must have washed down from the gorge, reasoned Jed while he noted the difference between gravel on the creek bed and the sandy banks. He stepped into the shallow stream, filled his coffee pot then placed it at the fire's edge. When the coffee had perked sufficiently, he poured a cupful. He sat down and rested beside the fire while, farther from the heat,

Bow slept.

A deer walked close to camp. The animal was exhausted and stumbled into a thick stand of cedars. Going over for a closer look, Jed found that the animal was almost unconscious. After placing apples and a pot of water beside the deer, Jed returned to the fireside in time to see Bow following the deer's back trail. Maybe I should have tied Bow, thought Jed as he started to pursue Bow along the creek bank. Something was obviously chasing that deer, noted Jed. The deer is completely worn out. The hunter must not be far away. I'll have to make sure Bow doesn't get into any trouble.

Although Jed could not keep Bow in sight, the dog's tracks were easy to follow. They mixed with deer prints and kept to the creek's bank winding through stands of cedars and old willows. Eventually the trail came to a lake covered by ice and snow.

When Jed reached the lake, he heard an outburst of furious snarling. The battle was brief and ended with Bow standing over a large hound. Growling continued. Teeth remained exposed and poised. Slowing, Bow moved off the defeated dog as Jed approached.

Seeing Jed, the hound cautiously got up and approached him. He reassuringly moved his hand along the side of the dog's head and said, "Hound, you go home."

Turning around hesitantly, remaining wary of Bow, the hound started walking back along his old tracks. I like hounds, reflected Jed. Like most dogs and cats, however, they do a lot of harm to wildlife. More should be done to control domestic cats. Society refuses to face the fact that cats are the worst killers of birds and small animals. The hound looked back once before moving out of sight among willows.

"Good work Bow," said Jed as he ran his hand along the dog's back. The dog's jaws seemed to form a smile in response to words of praise. He was clearly pleased with his victory and accustomed to being victorious. He and Jed returned to camp.

Jed checked the deer and found that the animal was still

sleeping. Bow showed only slight interest toward the sleeping animal.

Returning to the campfire, Jed poured himself some coffee then sat down, leaning his back against a pine trunk. He noticed the straight line of fox prints in the snow. With Bow, he followed these prints and came to an overhanging willow branch amid a large scattering of chicken feathers. "Our old friend the fox is continuing to catch chickens," said Jed to Bow while the dog sniffed at numerous fox prints. "That coop must be getting empty. There must be more than one coop where a fox can get chickens."

Jed prepared forked stakes. He stuck them into exposed earth next to the fire. On the forked stakes, a spit was placed containing moose meat. It sizzled while roasting and soon became a delicious meal for Jed and Bow.

After the meal, Jed checked the deer. The water pot was empty. Apples had been eaten. The deer had walked away leaving tracks that headed into the swamp.

Jed added more wood to the fire. Flames leaped upward, heating a wide area. Beyond overhanging branches, a shill cry of an hunting hawk rang clearly through the forest. Evening was approaching, taking warmth from the day. Fading rays of sunlight shone on ice coating tree trunks from the recent freezing rain.

Jed left camp to gather firewood in the swamp. Watery areas were numerous and often concealed by a covering of wet snow. With a startling crash, a slab of ice shot upright as Jed's snowshoes descended into black water. He toppled into the mire. He was able to reach a sapling and pull on it to help himself roll onto solid ice. When he stood up, icy water rushed along his skin and gathered in his boots. Water squished loudly with each step as he hastened back to camp.

He used two sticks to carry hot stones from the fire to the interior of the sweat lodge. His clothes were hung beside the fire before he carried a pot of water into the lodge. Water trickled onto hot stones sent steam billowing throughout the small chamber filling it with intense heat. Feeling the cleansing sweat, Jed exclaimed to himself, I'm warm again—and clean too.

When the steam became stiflingly hot, he left the lodge and splashed into the creek. He washed away sweat by using some ashes and sand along with water. While resting in the soothing, numbing creek, he noticed that a particular slant of the evening rays of sunlight shot golden wedges through the water and outlined pebbles on the creekbed. Intrigued by such colors and forms, he picked up a golden pebble. It was unusually heavy. His teeth could put dents in the pebble. It can't be, he told himself. It does, though, sure look like gold to me. It has washed down from the gorge. The storekeeper knew there was more gold to be found in this region.

The more he looked at the pebble and examined it the more he realized the discovery he had made or been led to find. The implications of such a find, preferably a localized and limited pocket, stirred his thoughts and he jumped from one possibility to another until he came to an height of exhilaration that he remembered others describing as fever—gold fever. He checked his wild imaginings, though, by cautioning himself that this was not likely his own work to be for just himself. He had, he reminded himself, prayed a bigger prayer and he was directed to this discovery because it was tied to the land and his bigger prayer.

The gold started his dreams soaring to the realization that he had found something larger than gold and grander than he could envision all at once. Here, in this creek, in the north woods, I have found not only gold but something much grander, he exclaimed to himself, standing in the water, holding the nugget and seeing his surroundings, and his life, for the first time, or, at least, in a new way. This is my larger prayer. I have suddenly realized that the messages my grandfather and the minister, Cal Tomkins told me are true. Wouldn't the world, and our interpretation of it, be different if everyone could know what I have learned that the traditional Indian belief and the Christian message are both real. These beliefs are not just legends. They are actual paths leading to the Creator, just as Cal and Tom said. After realizing the messages are real, discovering the bigger nuggets, we are, the rest of the time, just adding on the details. Hearing about something is one thing but believing it is an whole different thing—an whole different story. Learning about the beliefs in the traditional village is information; but believing them brings us to life itself.

The cold water was starting to jab him. He rushed to shore, put on dry clothes then returned to the stream's bed. He looked for gold and even used his frying pan to pan for the heavy nuggets until evening left him with insufficient light. From one place in the creek, he had acquired a bag on nuggets.

The next morning, aided by its sunlight, Jed continued his search. He found just one additional nugget. There had been only one pocket of gold, he concluded. There are no others, apparently. Judging by the weight of this bag of gold, I need no more anyway. I'll have the Heron Cove community purchase the entire logging company. Heron Cove owns this land. Payment for more lawyers will help secure legal, land title. The community will buy the logging company and put an end to destructive logging practices. This money from the land will also go into establishing a park to protect the place from where real gold comes. All people can come here to enjoy the land. It will no longer be destroyed for the profits of a few. There is enough money here, I think, to help Heron Cove carry out other programs that provide both employment and an healthy environment. I will just invest enough money for my own income. I am very thankful that Cal Tomkins and Tom Speaker provided good advice.

Additional searches yielded no more gold, confirming Jed's conclusion that he had found an isolated pocket. The excitement of his discoveries kept him from sleeping well. He packed his equipment and started traveling again with Bow.

They proceeded northward, following elevated ridges while crossing the swamp. Valleys were marked by deer trails. Porcupine paths led from one tree to another. Rabbit tracks were plentiful. After leaving this swamp, Jed and Bow came to the top of an hill overlooking the small lake. Jed selected a northern route that was more to the west than his previous journeys. Most of the forest was still and quiet. Little moved near the travelers except a shadow of an hunting owl flying overhead in a murky sky.

I have a general idea of my location, noted Jed. I'm north of Clarksville, east of Sky Lake and South of Heron Cove. There is a main road joining these three locations. I should also come to a secondary road before I get to Heron Cove.

The northward journey was measured by a string of campsites that marked a route through black spruce forests with rock cliffs and encircling blue hills. The travelers always seemed to have the company of ravens. Their resonant, croaking calls were a welcome sound as they rang through the wilderness and usually greeted each new day. Amid golden tints of sunlight, the birds frolicked above the forest while their large wings brushed a crisp, winter sky. From an hill, Jed finally saw the rooftops of the Heron Cove village. Sunlight emblazoned the forest with shafts of gold before turning to a red orb descending into a crimson horizon.

Fortunately, the village does not appear to have changed much, thought Jed as he served fried trout fillets to himself and Bow. Bow ate the food quickly then went to the open part of a creek for a drink. To his fillets, Jed added vinegar and pepper then savored the tender, juicy meat. After this meal, he sat down to enjoy a cup of coffee while he sifted through a flood of memories stirred by the sight of Heron Cove. I like things that don't keep shifting, he reflected. When things don't vary, they become dependable—reliable—and can be held on to like traditions. The same thing is applicable to people. I like people who are not always changing. In the morning, I'll go into the village. I'll convert the gold into money. I'll share with family and friends, especially including my grandfather and Cal Tomkins for his church as they showed me the way. I must buy an house and give money to the council so they can purchase the logging company without delay. Other projects can also be financed that will be good for the community along with the forest. We need ways of employment—preferably including traditional ways—that do not harm our environment. I know of paper companies that brought much more harm to people than employment because mercury polluted their water supply. We would be much healthier and more prosperous without such destructive employment. Caleb Pine's work in cleaning up Clarksville inspired me to revive my own life. What I saw Caleb do for Clarksville and its people I knew should be achieved for myself. The revival of Clarksville started my journey; so my personal road of recovery will be complete when I bring some benefits back to Caleb. We'll establish a park maybe centered at Caleb's Bear Trap Mountain. I'm going to miss my journey north with Bow. We must travel again whenever possible. I'm interested in what lies ahead, although I'm sorry my present journey has been completed.

In morning sunlight, Jed and Bow walked to the council house. I want to visit my family and friends. First, though, I must look after some responsibilities. He and Bow entered the council building, an elaborate structure beautifully made of logs. An hallway led to a receptionist's office. "Hi," said Jed to a woman who looked up from a computer she had been checking. A beaded hairpin tied her long, black hair away from an attractive face with dark eyes.

"Good morning," she replied smiling.

"When will the council be meeting," he asked.

"Tomorrow afternoon," she said, looking back at the computer obviously wanting to get back to her work.

"Is there any way of having them meet now?" he pushed.

"I don't think so," she replied, continuing to show little or no interest in this interruption.

"I have something very important to tell them," he stated.

"Who are you?" she asked.

"Jedediah Speaker," he replied. "Tom Speaker's grandson."

Eyes flashing, she exclaimed, "I heard his grandson was

coming back. I'm Sally Fiddler."

"Good to meet you," he replied. "Will you call the council?"

"No," she said, smiling again. "It would have to be important for me to start interfering around here. I like working here. I'll let everyone else try to be chief. But the receptionist is not supposed to try telling people what to do."

"What if it's a mistake to not call them?" he pushed.

"That's a chance I'd sooner take," she answered thoughtfully.

"Would you call Tom Speaker and, if he's home, ask him if he would assemble the council?" asked Jed.

"That I could do and right now," she exclaimed, picking up the phone. "You can wait in the council room," she added, pointing to the end of the hall.

"Thanks," he said. "The right person is in charge around here."

"Glad you understand," she replied, eyes flashing again.

Jed and Bow walked to a spacious room with a front platform containing tables and chairs. This main section faced an area filled with rows of chairs. Jed walked to a chair at the side of this room. Bow rested near the wall and soon was asleep.

In a short time, members of the council entered the room and either spoke or nodded to Jed before taking places along the front platform. Tom Speaker arrived and sat beside Jed. "Is this going to be a good meeting?" asked Tom.

"I hope so," answered Jed.

"Same here," exclaimed the old man whose face lit up with amusement.

Watching his grandfather, Jed realized that the old man had such depth of character and years of experience that the storms of life no longer shook him much and the possibility of calling a botched council meeting was something he could look upon with some amusement. Jed also appreciated the confidence Tom had in him in order to immediately call a meeting when requested. "Thanks for trusting me," said Jed.

"I've known you for a long time, Jed." Said the old man calmly.

I don't know any of the people on the platform, thought Jed. They seem comfortable and have that official attitude portraying their consideration that this meeting is just another customary part of what they do all the time. They talked to each other and surprised Jed when he heard his name called.

When Jed stood up, Tom whispered, "This better be good."

"Thank you for coming here on such short notice," said Jed to the councilors just as the receptionist entered the room. She carried a large tray containing cups filled with coffee. Each person, including Jed, received coffee before he continued to say, "I've been away a long time." His tone sounded so formal, he added, "Not that anybody cares." Laughter broke away the formalities, and the coffee helped too, then Jed said, "I've had a long journey back. I walked with my dog, Bow. My main journey, however, was not by foot; but came about by receiving advice from Tom Speaker and Cal Tomkins."

"That can be risky," interjected a councilor. A murmur of laughter followed before Jed said, "But I did not come here to talk about myself."

"Good," said a councilor and there was more laughter.

"I didn't even come here to talk about Tom or Cal."

"Good," interjected another councilor and laughter started again.

"Tom and Cal showed me the way to the Creator who showed me a gift I have to give to you."

"Now that is good," exclaimed a councilor and there were smiles all around.

"This gift," said Jed, "is for very specific uses. First, it is given to purchase—or buy off—the logging company that is clear-cutting trees south of here on Tom's old trapping area."

The councilors turned to look at Jed more intently as did Tom Speaker. One of the councilors said, "This must be some gift."

"The second portion of this gift," resumed Jed, "is for programs that help the community with such things as employment and traditions. This will include establishing a park to protect this land from here to Bear Trap Mountain. These uses must be strictly maintained particularly with regard to buying the logging company and protecting the land with a park. I have kept one portion to purchase an house as well as look after financial security. The main part is in this leather bag of gold nuggets."

Jed carried the bag to the front table. Each person checked the nuggets. The receptionist arrived with more coffee and said, "I know when to call a meeting." There followed a blend of joking, consternation and excitement in the room. "That Jed has really done it," shouted Tom who was aglow with happiness as were the others. "We can carry out our programs with this help." Turning more directly to Jed, Tom said, "You have done well. You have received a vision and carried it out."

"Thank you for your help," replied Jed. "Who is the chief here?"

"I'm supposed to be," answered Tom.

"Oh, I didn't know or I forgot," explained Jed. "It is most important to stop the loggers."

"That will be done first," answered Tom. "There are other threats to the land too such as the mining and oil companies. It isn't what is done so much as the way it's done that makes all the difference. We need jobs and a clean environment—not one or the other." Turning to all the people in the room, Tom said, "We can say that these nuggets have been kept from the earliest time in order to maintain the land and our traditions. This gold has been kept for us. It is part of the land and for the land. We have been with this land for so long we look back to no other homeland across the seas. We know there is no such thing as a coincidence. We were directed to locate and make use of this gold now, at this time, for the traditional prosperity of our people and all the people who will enjoy a forest that has not been cut down or polluted."

After a councilor had locked the nuggets in a vault, Tom said to Jed, "Come on, I have an house to show you and Bow." The two men and dog stepped outside. Other councilors were leaving in cars and trucks. Tom, Jed and Bow walked westward through the village. "You know," said Tom, "after we get this village back on its feet, you, Cal and I will have to do something about the pollution problems on a wider scale." When a scream was heard overhead, the two men looked up into a light blue sky to see sunlight flash on feathers of a soaring, circling eagle. "Yes, that's right," said Tom. "We are never alone."

# PART THREE

Jake Sand's appearance revealed his easy drift through life. He was moderately overweight because he enjoyed good food and rested much. There were no worry lines on his face although his hairline was receding and he was losing hair at the back. His face was somewhat plain and he wore glasses. He was of medium height. His easy, actually indifferent, slide through life gave him an edge of ruthlessness that enabled him, like fat, or oil, to move upward to become an executive of an oil company, and one of the companies, accused, before the cleanup, of polluting the Sand River in Clarksville. He was concerned about profit, not pollution, or any thing else getting in the way of profits. As one of the other executives had said, "We don't drink water." The cleanup of Clarksville put him, and his other executives out of work, although they were able to retire to a life of luxury. So now his easy lifestyle could continue without the interruption of having to go to the office and work with colleagues he didn't like anyway.

On a couch, in the living room of his house, Jake was resting, listening to robins singing in the first light of dawn when his large, front window exploded into an inward crashing blast of falling glass. A barrage of bullets kicked out glass remaining at the top of the window and tore plaster from the back wall, just below the ceiling.

At the first blast, Jake rolled onto the floor and crawled to a wall away from the shattering window. For protection from falling glass he scrambled under a fur rug on the floor. Bullets continued to enter through the broken window. They chipped pieces from mounted, animal heads that were hunting trophies on the back wall.

The people doing the shooting are firing too high to be trying to hit me, reasoned Jake before he used his hand to wipe sweat from his face. They are just trying to scare me—or have a grudge against me. Maybe they just have a weird sense of humor—or don't like my hunting trophies. My guns are on the rack on the east wall. I can't get to them.

By pushing the rug over pieces of glass on the floor, Jake crawled to a wall by the front door. He proceeded to the kitchen in the southeast part of his house. Cautiously, he peered over the sill of this room's one window.

Amid outside shapes, he noted the familiar outline of his tall, blue spruce. Gulls circled overhead in a gray sky. Robins had stopped singing. The dawn was quiet and still.

The spruce adorned a front lawn bordered in the east by an highway. Beyond this road, fields extended to misty, blue hills. Two people stepped out of a ditch and walked southward on the far shoulder of the road. They carried rifles. I can't identify one of the people because they are too far away, noted Jake. The other person is unmistakable. She's slim and has a strong, upright walk. That's Catrina all right. She turned out to be a strange wife. We've been separated for three years. Now she has started shooting at me. I wonder what's bothering her.

Songs of robins filled the dawn again. Sea gulls called from overhead. First sunlight of the dawn highlighted the countryside with pink hues. Preparing coffee in the kitchen, Jake thought, my wife and I have drifted a long way apart. I'm a successful executive and she's some kind of an environmental extremist. The environment came between us. We argued about it all the time, particularly after I moved up in the oil business. I can't believe she'd resort to shooting at my house. Although the shots were high—not really trying to hit me, they did a lot o' damage.

A cardinal's song entered through the screened window. Carrying a mug of coffee, Jake opened a door at the back of the kitchen and stepped onto a patio. The floor was constructed of colored tiles held in place by cement. Jake sat down on a wooden, rocking chair and faced the west. A table beside him held a box of shotgun shells. He placed the coffee mug on the table then loaded a gun that had been leaning against the side of the house.

He sighted along the long, well-oiled barrel until it was in line with a starling on a fence post. The gunshot seemed particularly loud in the confines of the patio and echoed among the hills. Chips were knocked from the battered post. Feathers drifted above the place where the bird had fallen. I'm keeping the starlings away, noted Jake. However, nothing else seems to be trying to nest here. I guess the birds don't know I only shoot some birds and not all of them. I would like to see bluebirds nest here again. I only seem to get starlings. I've shot twenty-six this spring. There still seems to be lots of them.

He leaned his shotgun against the wall before he drank more coffee. If I don't like something, I just try to get rid of it, reflected Jake. I've been somewhat active against things and people I haven't liked to the extent that nothing has bothered me—until today.

Crows flew to fence posts farther back on the property. When Jake reached for his gun, the birds left the posts, flew over a field, and found refuge in the forest. Crows survive because they are smart, observed Jake as he replaced the gun. Being smart has also saved me a lot o' trouble and work. Life has been rather easy so far. I made a lot o' fast money in oil and mines until Caleb Pine started cleaning up Clarksville and I, along with others, had to adapt or go out of business. I almost had enough money to retire. I don't understand why the environmentalists are such extremists. I'm not fanatical about oil or mines. There's a lot o' fast money to be made as long as we don't let the tree-huggers and water-huggers cut into our profits like what happened in Clarksville.

Jake drank the rest of his coffee then returned to the kitchen. Dishes seem to have rusted to the sink, he noted while placing his cup beside a pile of pans and plates. I should have put them in the dishwasher a long time ago. This place is very dusty and I've seen mice for the first time. I might have to hire someone to do some cleaning.

Jake left the house, removed his car from the garage and

drove south near Clarksville where a construction company had come back to life with all the new activity in Clarksville and surrounding area. Leaving his car on the parking lot, he opened a door in front of the building and stepped into an office. "George Willis," exclaimed Jake, boisterously to an employee who had looked after construction projects for Jake in the past. The man was slim, short and balding with neatly trimmed, gray hair clustered around the sides and back of his head. The glasses he wore gave a sparkle to a wrinkled, friendly face with gray eyes. "You've done a lot o' good work for me in the past," continued Jake.

"Yes," said George, congenially, "and you've always paid well in the past."

"And intend to again," countered Jake. "Have another job for you if you have time."

"During working hours-or after," said George.

"Thanks," replied Jake. "My wife visited me this morning."

"Oh," said George, pensively. "Or is that an oh-oh?"

"More like an oh-oh," answered Jake, laughing. "She shot out the front window of my house."

"That's more like a wow," exclaimed George, "and a little unusual isn't it?"

"There was a time when I would have thought so," answered Jake. "Please take a work crew to my home and replace the window."

"When Clarksville came back to life, we did too," said George. "Your window, though, is sort o' an emergency. We'll work that in right away even if it's after usual hours. We recently updated all your windows. The front is a standard size. I can get one. We'll work on that right away even if it's after the shop closes—today."

"Thanks for the extra help," said Jake. "I'll see you out there," he added, turning away and leaving the shop.

Having not much he could do until his house was repaired, he decided to enjoy the rest of the day by going for a long drive and used his phone to arrange for people to come over for a spontaneous party. Instead of an house-warming party, he would have an house-shooting party. He lastly called George to inform him and invite him along with his crew. Having made the calls, his thoughts returned to the better days he had enjoyed with his wife. Like most things in my life, I met Catrina by accident, he reflected. I'm not a fisherman and not an outdoorsman. I thought the fly-in fishing trip would be a change as well as a rest. Of the people on that trip, Catrina was the most helpful in showing me how to fish and cope with the wilderness. I guess she thought I had learned more than I actually had. After we got married, the differences became more obvious until we clashed about almost everything—particularly the environment.

Ready for an evening meal, he stopped at a restaurant to enjoy some chili con carne and beer. He had too much beer and was not thinking clearly before the restaurant seemed to start moving around him. I can't drive now, he told himself. I'll call a taxi and leave my car at this restaurant.

"Someone is having a party up ahead," said the driver to Jake as the taxi went past cars parked on the side of the road. Loud music throbbed into the night and came from an house partially hidden by a stately spruce.

"This is my place," replied Jake who vaguely remembered inviting his usual party crowd to his shot-up house. "I guess they started without me. They usually do anyway."

Jake fumbled with his money and paid the driver. He left the cab and walked past parked cars on the way to his house. The front window had been replaced, he noted with satisfaction.

Opening the front door, he was met by pungent aromas of smoke, food and alcohol. They revived a feeling of nausea. Lights were bright and music loud.

Although Jake's thoughts were blurred, he felt a rush of anger when he was knocked backwards by a man who had been punched by another. Jake grabbed the shirt of each fighter, twisted the material near their throats and choked them enough to be able to lead them outside. Before falling over some juniper bushes, one of the men punched Jake on the side of his head, sending him groggily back inside.

He reached a bar that had been set up in his living room. Someone gave him a drink. He sipped it and discovered it was rum mixed with coke. Much of this drink spilled when he pushed through a dancing area and walked outside to the swimming pool at the north end of his house. There are a lot o' people here, he thought, and I don't seem to know anyone. A lot o' people are around the pool although only two of them are swimming. Deciding suddenly to return to his house, he bumped into a woman who yelled, lost her balance and fell into the pool amid a background of laughter. Someone else was pushed into the opposite end of the pool. Jake was shoved from behind and he tumbled into the water. He thrashed wildly, desperately struggling to breathe. When his arm hit the pool's ladder, he pulled himself onto the tiled deck. He choked violently.

Regaining normal breathing, he rolled under some juniper bushes and was soon asleep. He woke up later and saw a starlit sky along with a pale moon. He slept again until being awakened by rain tapping against his face.

Shaking with cold, he stood up and walked inside his house. It smelled of smoke, stale food and alcohol. The place stinks; but someone has cleaned up most of the mess, observed Jake. Must have been George Willis. Even my wife liked him. She said he and Jack Kelsey were the only good characters in my gang of cronies.

Following a shower and some coffee, the morning looked better to Jake. Rain tapped against the windows. Robins' songs rang clearly through damp air. Crows cawed in the distance.

Jake carried a cup of coffee upstairs to his new office. When he had been forced out of Clarksville because his company had been polluting the environment, he was able to leave with enough money to almost retire comfortably and have time for his new project displayed in his attic. A cot was located next to the west wall. A kitchenette and washroom were to the north. A large window faced southward. Looking out this window, Jake noted that the morning was brightening over fields reaching to a willow swamp bordering the forest.

At the base of the window, there was a desk containing two lamps. A chair was next to this desk. A large table, surrounded by chairs, filled the center of the room. This table was covered by an elaborate model of an housing, development plan titled Sands Estates.

Jake turned his attention to his model. On its western border, there was a replica of his house. In front of it, a painted strip marked the highway. To the southeast, across the highway, there was the naturalists' cabin where Catrina worked. Adjacent to this property, there was his friend, Jack Kelsey's cabin.

The line, marking the highway, went north and came close to Sky Lake. A gravel road went to the lake. Here Jake noted the miniature of his cottage.

Bringing his attention back to the model of his house, he checked the area across the highway. First there was a field then a wilderness area stretching northward past his cottage and including Heron Cove to the east and Bear Trap Mountain to the northwest.

Eastward from Jake's house, across the highway, was located the beginning of his plan. I don't know why my wife and the other naturalists are so opposed to my idea of developing vacant land. The land isn't being used now. It's a wilderness. There's a claim against my ownership coming from Heron Cove. That's why I pay a lawyer. You can't make money from unused land—or unused trees. I have planned for maximum use of space to increase profits. Sands Estates will provide prestige homes as well as condominiums.

Jake noted the location of each painted road as it wound among miniature stacks of condominiums mixed with individual houses and entertainment facilities including tennis courts, swimming pools and golf courses. Each neighborhood will be fenced for security, observed Jake. The streams will flow to fishponds. Everything will be provided for easy living. I've hired a local developer, Max Coker. He was hard to get because things are booming around Clarksville. He is to start the first road right away. That'll set things in motion. That's what I like—lots o' action—as long as it's other people I get active and I don't have to do too much myself except make plans and lots o' money.

While outside, the sky was clouded and rain pelted the windows, Jake rested on a chair in front of his fireplace in his living room. He enjoyed the warmth and sound of the flames. Opening the road is essential, he told himself. Once we start cutting trees to clear the roadway then everything else will follow. Max said he would have a man start cutting trees today. The road will leave the highway in front of my house then go straight to the forest. This is a gray, cold day. I can help shake off dreariness by getting the road started. Having removed trees to make a route for the road, we will start clear-cutting the forest and sell the timber. After I have cut the first tree, I can also find out what the naturalists plan to do, if anything. There isn't much the tree huggers can do. They can enjoy hugging logs. I have made some well-placed political contributions. I bought the government owned land that includes the forest on the understanding I would develop the area. Governments like taxes. We need employment more than trees. My wife says the forest doesn't pay taxes and also doesn't charge for the oxygen supplied for us to breathe or clean water for us to drink. She says I'm destroying the real value of the area in order to make a quick profit. If I make money now and others are employed now, I don't care if no one is employed in the future. I will have enough money to go somewhere else if necessary.

Jake poured rum into a tall glass. Next he added coke followed by ice. The work must start today, he concluded. After the first tree has been cut, I'll begin to make a lot of money. I just have to get the work started. Jake drank the rum while enjoying heat from the fire.

Just as the rum started to flood his thoughts, there was a loud knock at the front door. He walked somewhat unsteadily to the door, opened it and found himself confronted by the lanky form of Jack Kelsey. His shoulder-length, black hair matched the color of his full beard and eyebrows. The eyes were grayish-blue. Lines had formed around his eyes as well as across his forehead. He wore a black jacket, cotton shirt and jeans along with leather boots. "Hello Jake," said the man after taking the cigar from his mouth. "I've been hired by Max Coker to start clearing a roadway. I was supposed to start today."

"Like some rum?" asked Jake.

"Thanks, but I'll have to wait 'til later," he said. "I have work to do."

"Come in," replied Jake. Jack entered the house. He followed Jake upstairs to see the model. Jake pointed to a road leaving the highway in front of a miniature of his house. "This is where you will be cutting," explained Jake as his finger followed the road. "Survey work has been done. Starting at the front of my house, the road goes east to the woods. There are stakes marking the route. There will be much cutting to do in the forest. Eventually, most of the trees will have to be removed. The road must be cut first. A larger, work crew with heavy equipment will arrive after you have started the work. You know the woods quite well, don't you?"

"I used to cut firewood in the forest before you bought it," replied Jack. "I shoot deer there also. With wood and deer from the forest, along with my own spring, I haven't had to travel much for food. I can get most things near home."

"You know what we are doing," said Jake. "You have lots of work to do. Maybe we should get started."

The two men walked to the doorway. Jake put on a jacket while Jack picked up a chainsaw and related equipment he had left on the steps. Air outside was cold and damp.

"How well do you get along with the naturalists next door to you?" asked Jake as he and Jack walked toward the highway. "By the way, if you had arrived any later, I would've had too much rum. The excitement of getting this project started seems to have cleared my head."

"You seem to be all right," observed Jack. "As for the naturalists, they don't like the fact that I shoot deer. Those people resent hunting because they feed the deer and most other things. I was hunting deer before naturalists moved to the house next door. They can mind their own business and I can mind mine."

After the two men had crossed the highway, Jake said, "The environmentalists are against you because you are an hunter. They are opposed to me because I was a polluter and now I'm developing the woods east of here. We'll see what we have to do. When the wind is blowing the right way, a grass fire would take care o' their cabin."

"Sometimes grass fires can be a problem," laughed Jack before returning a cigar to his mouth. Striking a match, he relit the cigar, sending a trail of smoke curling away from his face.

Noting sparsely wooded fields leading to the forest, Jake said, "Loggers first cleared fields near the highway. Settlers tried to farm this land about an hundred years ago. Farms petered out because the soil is sandy as well as rocky. Before the loggers and settlers, the explorers arrived then traders, missionaries and trappers. There are the remains of a trapper's cabin in the forest behind my house. Apparently, the first people to live here were Algonkians who came from the east and north. The Iroquoians arrived from the south. The Iroquoians, who came as far north as the Clarksville site, were Mohawks who traded with the English. Meanwhile, the Algonkians traded with the French. All these visitors have come to the forest that we see from this field. The forest hasn't changed. It holds many stories."

"I haven't been in the forest much," observed Jack. "I've found the hunting to be adequate in these fields—and sometimes out my back window."

"I haven't even been in the fields much except to check this road," admitted Jake. "A survey crew put in stakes for the road. Some people have gone into the woods and not come out. Apparently, it's a wild place with swamps and quicksand. I guess people would also just get lost. This forest, itself, has been subjected to very little logging."

"Stories help keep people away," said Jack. "Glen Kirson looked for gold in there. He just disappeared. Other prospectors vanished also along with hunters. I heard that someone found two skeletons that had the heads missing. Kirson's wife vanished in the woods before he became lost himself. There are the usual rumors that people have heard Kirson's wife calling at night. Some hunters were camped at the edge of the woods and her ghost walked through their camp at night. The ghost became smaller and brighter before disappearing. There are wolves in the woods too. I've heard them howling at night. I got lots o' reasons for staying out o' the woods."

"Except to cut a road," added Jake.

"A road is not a woods," countered Jack.

Hearing a cry from overhead, both men looked up and saw an hawk circling through a gray sky. Bringing their attention back to the land, Jake pointed to a stake and explained, "From that stake to that distant spruce, the road is clearly marked and there are no trees to remove. The spruce is the first tree to be cut. There are others on the way to the forest where the real work will begin. We'll see how many mysteries there are when we get to the forest."

"Am I supposed to work alone?" asked Jack.

"You will be followed by a work crew using heavy equipment," replied Jake. "The main aim of today's effort is to begin our project."

A shadowy form of an hawk flew past Jake and Jack. The bird veered into tall grass and a rabbit squealed. "There has been a family of hawks around here for as long as I've been at my house," noted Jack. "There must be a nesting site in the forest. A lot o' owls are around also along with some vultures. Noticed a few cormorants flying by quickly in the last year or two. The warming climate brought them. I shot a vulture a few days ago. The bird was gliding slowly above my house. I brought the bird down with one shot from my rifle. The first shot broke a wing. A second shot was needed. Both shots brought some shouts from the naturalists."

The men adjusted their jackets to keep out an edge of a cold breeze bringing more rain. Pointing through the rain, Jake said, "Maybe we could at least knock down that spruce today in order to begin. Just starting any job seems to be the hardest part."

"Did you have any trouble buying this land from the government?" asked Jack who was reluctant to get working in the rain.

"The government's representatives knew that my development would pay a lot o' taxes," answered Jake. "The only real opposition has come from my wife and the other naturalists. They say I'm turning the area into a future wasteland for a quick, short-term profit. I think this area is a wasteland now. I don't think the environmentalists have the money or influence to stop me. I have used political influence, law and money to get what I want. I don't know why the environmentalists always seem to have slight, political influence, insignificant legal affect and little money. I was lucky acquiring money. After I became rich, everything else has been quite easy."

"I've had the opposite type of experience," replied Jack. "Since I've never had much money, most things have been quite difficult. Being an hermit comes naturally when you're broke."

"I'm kind of an hermit and I'm rich," countered Jake.

"Yes, but you have a choice," stated Jack.

"I think my wife has decided that I should be an hermit," laughed Jake.

"My poverty decided that I would be an hermit," said Jack.

"You'd get richer if you'd go and cut down that tree," said Jake.

"I'm going," replied Jack. He took his saw and walked eastward beside stakes leading toward the distant spruce.

Jake decided to accompany Jack and soon caught up with him. "If the naturalists were smarter, they would have purchased this land—like I did," said Jake. "They always seem to be following the developers' advance."

Without replying, Jack kept walking, having decided to get some work done. The two men crossed rolling land containing a few maples as well as elms. Developing leaves added a greenish hue to the trees. Trout lilies were numerous along with trilliums. Hillsides rippled with spring grass. Mosquitoes and some black flies harassed the two men while they approached a tall, white spruce, the first tree blocking their route. "The road could go around this tree," admitted Jake. "Removing it, though, is easier than turning the road."

"I'll have the spruce down in a few minutes," replied Jack. He started his saw and cut off lower branches in order to have clear access to the trunk.

By making an horizontal cut into the trunk then angling an upper cut down to meet the first, he removed a chunk of wood, leaving a notch in the tree. Moving around to the opposite side, he started a third cut which was almost horizontal yet angling downward somewhat toward the point of the notch. The trunk cracked before the top of the tree moved, gathered speed and toppled with a whooshing crash. Branches were afterward cut away from the trunk before the saw stopped and silence returned to the fields. "As maybe you know," explained Jack, "the first notching cuts determine the direction the tree will fall —without wind interference or weight distribution factors of course."

A screaming sound of a bullet whooshing over their heads, followed immediately by a distant gunshot, sent Jake and Jack scrambling for cover beside the tree's fallen trunk. Diminishing echoes were shattered by a second shot. Both blasts came from the vicinity of the naturalists' cabin. Silence had almost returned when more shots screamed overhead.

After the firing had stopped, Jack said, "The shots were a long way over our heads."

"My wife and her friends were doing the shooting," replied Jake.

"That's a strange wife you have," stated Jack. "I'm pleased

that I'm not married. Naturalists are all the same. They're extremists."

"My wife will come to see the error of her ways," said Jake, confidently.

"You think she will change?" asked Jack.

"She has no choice," stated Jake. "She can't stop my project. They are lucky I'm not reporting their last attack—or this incident."

"Maybe we've done enough work for one day," offered Jack.

"I don't see how people can care so much about one tree," observed Jake.

The two men left the trunk and walked back to the highway. Jack returned to his cabin while Jake went to his house and started a fire in the fireplace. A clouded sky increased the darkness of the night.

When the clouds scattered, a full moon cast silver light upon two people walking toward Jake's house. "Jake doesn't work very much," said Catrina to her friend, Madeline Cooper. Madeline was a fashion conscious person who kept a gray streak in her curled, black hair. Her eyes were also black as was her jacket. It highlighted a white blouse she wore along with designer jeans and western boots. Catrina also wore western boots and jeans along with a red, cotton shirt beneath a sheepskin jacket. Her eyes were deep blue. Straight, black hair bordered her tanned face. "Jake doesn't think very deeply about anything," added Catrina. "He's a dilettante—a trifler. We should be able to stop his project because, if it becomes strenuous, he'll quit."

"I have noticed that you are not a quitter," stated Madeline. "I'm a city person who respects the wilderness. You are a country person who can live in the woods. You can even ride horses. Although I like horses, I have no great interest in riding them. Considering your interests, I'm surprised that you'd marry Jake Sand."

"When I told him about the wilderness," explained Catrina, "he listened politely—or, at least, he didn't say anything. I thought I'd given him a new way of life, and seen something in him that I liked. Later, it appeared that I had wasted my time and hadn't reach him at all."

"Fortunately, he had no affect on you either," added Madeline.

"We drifted apart quickly," said Catrina. "Maybe I'm still trying to get his attention."

"He must be very hard to reach," joked Madeline. A smile brightened her face.

"I'm not just trying to get his attention," continued Catrina. "We have to stop him from wrecking the forest. The wilderness is far too beautiful to be turned into a quick-profit, moneymaking scheme for the benefit of Jake Sand or anyone else. People like Jake protect themselves with walls of contentment based on money along with legal or political influence. Behind these barriers, such people can't hear, and don't want to hear, a view that is different. Jake has a lot to learn. He, at least, deserves help."

"Maybe you should give up trying to change Jake," suggested Madeline.

"I could give up on him," Catrina replied. "However, I won't surrender the forest to his development plans. We didn't get rid o' the polluters and environment destroyers in Clarksville just to have them continue their destruction in different places—and, particularly, not out here."

The night was quiet. Clouds intermittently scattered, allowing moonlight to brighten the landscape. Wolves howled in the eastern forest.

The two women walked to the lofty, blue spruce in front of Jake's house. After lifting a lower branch, they crawled inside the canopy and moved to the trunk. Occasionally, through a covering of boughs, moonlight entered, providing patches of light while Catrina's bow saw made a first cut on the side of the trunk that faced Jake's house. A second cut met the first and a chunk of wood was removed to form a neatly sliced notch. On the opposite side of the trunk, a third cut was started and angled down toward the point of the notch. A resinous sent of spruce pitch filled the cramped space under the branches where the women worked. "Are you sure this tree will fall away from us and go toward the house?" asked Madeline.

"I think I'm sure," answered Catrina. "When the trunk cracks, well get out o' here anyway. The blade keeps sticking in the pitch. Oiling the blade keeps the saw working. There's no wind. The tree will fall the way we aimed the notch."

"I'm sorry we are cutting down such a beautiful tree," observed Madeline.

"I'm sorry too," replied Catrina. "But maybe Jake will come to see how we feel about a forest."

"That's a long shot," said Madeline. "We should finish this work and get out o' here."

The hooting of an owl seemed particularly loud. The sawing continued until the trunk cracked. When the tree started toppling toward the house, the two women scrambled out from under lifting branches and bolted toward the highway. Behind them a shattering crash broke the night's solitude. Looking back, the women saw the long, dark form of the spruce imbedded in the front of Jake's house.

An explosion of wood and glass sent Jake tumbling off his couch. Wedging himself next to it for protection, he waited and listened. Tree branches were everywhere—and glass. What the hell was that? Jake asked himself. A tree is right in my house. A tree came through my house. Must've been the big spruce on my lawn—my tree. We must've had a storm.

Pushing slowly amid branches and being careful to avoid pieces of glass, he reached a clear place and turned on a light. A gaping hole remained where his new front window had been. The window area was now filled with a massive trunk and branches. I must be a sound sleeper to have slept through a storm like that, he told himself. If I'm going to miss things like this, maybe I should stop drinking, or cut down a bit anyway. This is a disaster—like the last time when my wife.... Surely, she wouldn't—not my spruce.

Jake went outside to see what had happened. In moonlight, he saw the white wood of the freshly sawed trunk. She wouldn't, he told himself. She wouldn't cut down a tree because she likes trees. Of course this was my tree—my house. She has gone too far this time. No one cuts a spruce tree from a person's lawn. What kind of a mind does that woman have?

Jake could not get back to sleep. His mind would not stop thinking about the wreckage of his house—and his tree—and his wife. Also, there was something moving in his house. Maybe a bat came in, he reasoned. Usually, I wouldn't hear a mouse; but I can hear this thing. There are mosquitoes in here too.

In the kitchen, he made coffee then sipped it slowly until sunlight brightened the morning. His thoughts swirled around his wife and the naturalists. Their cabin is beside Jack's home, he told himself. I'll call Jack and ask him if anyone is at the cabin. If no one is there, I'll call Max Coker and have him immediately move that cabin along the staked roadway, beyond the place where Jack and I cut down the tree. The work crews could use such a building. I won't bother to mention that the structure belongs to the naturalists. After all the environmentalists have done to me, I don't think they'll report the loss of their cabin. With one move, I might get rid of the naturalists and also get a shelter for the workers.

After calling Max and Jack, Jake took an extra cup of coffee with him when he left his house and drove to the construction company to see George Willis. When Jake entered the building, George looked up from an order desk and Jake said, "Thanks for cleaning up the house—both times."

"The window crew helped," replied George. "We couldn't leave a mess like that for you to wake up to in the morning."

"The others had no trouble leaving a mess behind," stated Jake. "Thanks for your help. You and your crew did so well the last time I was wondering if you would come out as soon as possible and put the window in again along with other repairs?"

"Again?" gasped George, incredulously. "Not your wife

again though?"

"You're right both times, George," exclaimed Jake. "The first time, she sent bullets flying through my window and this time she cut down the blue spruce on my front lawn and dropped it neatly right through my front window and did other damage too. My house is now filled with a tree along with any critter that needs a home. If you could fix this mess as soon as possible, it would be greatly appreciated—again."

"That's one woman I'd never want to see riled," exclaimed George. "If she gets riled, I hope I'm never one of the polecats she puts her sites on. Oh, and I keep your window in stock. Seems to be a top seller."

"I don't want to interfere with your business," replied Jake, "but I intend to take care of the environmentalists. I've asked Max Coker to have his workers move the naturalists' house across the field to the edge of the forest where I'm building my road. The road crew will need a place to stay."

"Did you tell Max you don't own the building?" asked George.

"He didn't ask about ownership and I forgot to mention it," answered Jake. "With all the work Max is doing for me, the moving of one cabin didn't concern him at all."

"The naturalists will be concerned," countered George.

"They won't say anything because of all the things that have been happening to me," replied Jake.

"I hope you know what you're doing Jake," said George, standing up, taking some work orders from his desk. "Your wife's been understanding so far but I sure wouldn't push her. I have lots o' work to do. I'll get a crew together—for our best customer. There's nothing we like better than a repeat customer. We get that feel-good feeling knowing they liked what we did the first time."

"You'll never be unemployed George," said Jake. "You could always be a comedian."

"If you keep giving me this material, I could do that easily," he countered with a smile brightening his face. He turned and started walking toward a back door.

"Thanks for your help," said Jake. He left the building and walked to his car. The morning was damp and cold under an overcast sky. He drove to Jack Kelsey's home.

Parking his car in Jack's driveway, Jake walked toward Jack who had been splitting wood. Jack said, "This should be an interesting day." As he spoke, both men heard sounds of heavy equipment approaching from the south.

Jake and Jack walked to the naturalists' cabin and helped direct the removal of the building. It was hauled to Jake's staked laneway and taken past the fallen spruce to the edge of the forest. Afterward, the work crew returned to the naturalists' land and, as much as possible, removed signs that a cabin had existed on that site. The workers finished the day by helping George Willis' crew repair Jake's house.

The next morning, Jake was in a good mood. He decided to enjoy a full day of rest by driving through the countryside. Apple trees containing pink and white clouds of blossoms bordered the roads. Fragrances drifted with air currents entering the car. Fiddlehead ferns reached from a forest floor mottled by patches of purple and white trilliums. Gold flowers of swamp marigolds carpeted numerous wetlands. Innumerable yellow patches of trout lilies patterned drier areas. Starlings nested in hollow fence posts and in some farm, mail boxes.

Amid the passage of scents and colors, Jake felt a rare sense of real contentment while he approached his house. Being in a restful mood, and driving somewhat habitually, he almost crashed into the naturalists' cabin that blocked his driveway and covered part of his front lawn. Bewilderment flashed to frustration then anger as he fully realized what had been done. Deeply rutted tire prints from heavy equipment crossed his lawn and led to the driveway.

"This can't be happening!" Jake shouted, getting out of his car. "They can't do this to me!" he stated angrily, slamming the car door. "They've recovered their cabin and parked it on my front lawn. Is there anything they won't do?"

The next morning, Jake telephoned Max Coker and asked him to get a cabin off his driveway and take the building to the back of his house then along a lane beside the fields to the edge of the forest. Jake also received a call from Jack Kelsey who said, "I called to say there was an house on your front lawn."

"Thanks Jack," replied Jake. "I hadn't noticed and might not have if you hadn't mentioned it."

"The environmentalists have been busy," continued Jack. "They have moved a larger clubhouse onto their property at the same site we just cleared."

"I don't want to see you get into any trouble," said Jake. "However, this is a bad time of year for grass fires."

"I expect there to be a fire here soon," answered Jack. "I would like to see the naturalists leave because they have caused a lot o' trouble."

"We shouldn't get actually mean about things," said Jake. "On the other hand since these people are naturalists, maybe they would enjoy one of your bag of tricks that I've known you to use before—and this is the right time of year too. We could consider it an house-warming. It's the only neighborly thing to do."

"I'll get the bag o' tricks right away and see you later," said Jack. He hastened to his cellar where he obtained a burlap bag.

From bushes near his house, Jack cut a pole. To one end, he secured a rope noose operated by having the other end of the rope extend to the pole's handle. My trick catcher is ready, he noted with satisfaction.

He walked to adjoining fields. He carried his pole and noose along with the sack. I hate snakes, he told himself. However, people say garter snakes are harmless. Maybe this stunt is a waste o' time; yet something has to work in order to get rid o' the naturalists. Moving their cabin failed. If they hate snakes as much as I do, maybe a little bag o' tricks will be successful.

Sunlight broke from an overcast sky and highlighted the yellow feathers of a meadowlark singing from the top of a fence post near Jack. A cardinal sang from the top of a maple.

Emerging leaves rustled in a breeze waving grass covering much of the fields. Flowers added traces of fragrances to the air. Enjoying the signs of spring, Jack came to an old stone foundation. He swung his pole to the base of a wall and managed to loop the noose around the head of a garter snake. A tug on the rope held the snake until it was put into the bag. Jack repeated this maneuver until the bag bulged.

Because there were no cars near the naturalists' house, Jack rushed forward. I must get the snakes in there and get out before anyone comes back, he warned himself.

Jack heard pounding sounds followed by a growl then saw a police dog, in full charge, coming around a corner. Behind the dog, a long leash raked the air. He's broken the rope, exclaimed Jack to himself. Although gripped by fear, he kept moving. He pulled off his jacket and held it forward. Jaws clamped onto the jacket. It was swung furiously from side to side.

While the dog shook the coat, Jack ran, carrying the bag. He reached the building's back door, opened it, threw the heavy bag inside and closed the door quickly. He turned and saw the teeth. The dog was attacking again, stepping forward slowly, issuing a deep-throated, guttural snarl. Jack tore off his shirt and threw it at the jaws. Teeth took the cloth and shook it viciously. Jack ran with all his strength. He reached his cabin, got inside and scrambled to shut the door. The dog's head came through the screen. The head retreated when the main, wooden door was closed. Jack rested, leaning against the door. He was sweating. His heart was pounding and his mouth was dry. We have to get rid o' the naturalists, he resolved. No one could tolerate neighbors like that.

Sounds of a car moving onto the driveway by the naturalists' cabin brought Jack's head above a windowsill of his house. He watched as Catrina Sand and Madeline Copper stepped out of the vehicle. They unloaded parcels of groceries before walking to the back door of the building where the dog greeted them. With some difficulty, they took a piece of shirt from the animal. Next they found another piece then a jacket. By reattaching the broken rope, the dog was again secured to a doghouse behind the building.

Madeline opened the back door and entered first, followed by Catrina. Jack laughed when the screaming started. The two women ran from the building and kept screaming. They deserve the snakes, thought Jack. Those women almost got me mauled by a dog.

Jack walked to Jake's house to report on the successful bag o' ticks. "I don't know what will be needed," said Jake, "to get rid of the naturalists. We have tried a more drastic move by taking away their cabin. That tactic didn't work. Maybe a smaller trick like a bag of snakes will work. Anyway, those people have got our message that they are not wanted around here."

"You should've seen those women running from their house," laughed Jack.

"I should've been there," replied Jake as he gave Jack a tall glass of beer.

"Thank you," said Jack, receiving the beer. "We won this battle, although that's a real, vicious, guard dog they have over there."

"My wife has always underestimated me," observed Jake. "Maybe now these people will give up and go away."

"Thank you for the beer," replied Jack. "I shouldn't leave my house undefended for too long. I must get back."

Jack walked to his house. He prepared a simple meal of biscuits and beans. Being tired, he decided to go to sleep early.

"He has just turned out the light," said Madeline with real excitement revealed in her voice as, from the window of the clubhouse, she and Catrina watched Jack's cabin.

"We won't have long to wait now," replied Catrina.

A blood-curdling scream came from Jack's shack. The lights came on. The back door opened before Jack threw something outside.

"Jack Kelsey is lucky that we are good people," said catrina to Madeline, "or we would have put a live snake in his bed."

"The light is still on," replied Madeline. "Maybe he is

suspicious and looking for more snakes."

Another loud yell tore from Jack's house before the back door opened again and something was thrown outside. "He got the one in the refrigerator," laughed Catrina.

"Maybe that was a bit mean although the snake in his bed was worse," said Madeline. "The poor man has had a bad day."

"His day has been worse than the one he gave us," said Catrina. "Maybe Jack and Jake will leave us alone now. They would quit if they knew us better. Jake never did take me—or anything else—seriously."

"He seems to be serious about his development plans for the woods," countered Madeline.

"He just wants to make more money so he can do nothing," said Catrina. "He would probably move to his cottage after he destroyed this area."

"Why did you marry him?" asked Madeline.

"I thought I had trained him," laughed Catrina. "I saw something in him at first—something that was good—and thought I could train him for the rest of it. He seemed interested when I told him about the wilderness. Maybe, he wasn't really listening."

"Jack has left his house," noted Madeline. "I guess the poor fellow can't sleep so he's going to visit Jake."

"If we stick to what we're doing, those two will eventually give up," said Catrina. "We can't let Jake have the forest."

After a short meeting with Jake, and more beer, Jack returned to his house. Sleep did not come to him that night.

For much of the next day, Jack sat in a chair behind his house. He watched birds carrying sticks and grass to nests being constructed. Crows robbed the grackles' nests. Hawks hunted birds and snakes.

Jack watched the naturalists build a fireplace behind their clubhouse. They first dug a shallow hole. It was then filled with sand and bordered by stones. The new fireplace was used for cooking as well as burning brush.

One, large fire burned throughout the night and was also

burning the following day when no one was present at the clubhouse. No one is over there, observed Jack. The dog is tied to the doghouse. Maybe I'll go over for a visit.

Jack walked to the environmentalists' cabin and knocked on the back door. Since there was no answer, he opened the door and stepped into a kitchen with an adjoining dining room. It, in turn, opened to a living room with bedrooms and washroom at the back. A cautious check of each room assured Jack that no one was sleeping anywhere.

Jack left the building. He noted again that the dog was tied to a doghouse and it was a safe distance from the clubhouse as well as the fireplace. A cardinal sang from a maple located behind the fireplace where a flame continued to flicker across some planks. Being able to hold ends of planks untouched by fire, Jake removed two burning pieces and helped the flame spread to the building, as if—possibly—an errant breeze had spread the fire. Although the dog snarled and barked, the rope held. Flames climbed quickly across dry wood on the building and needed no help to quickly engulf the structure, sending a plume of smoke, like a living thing, curling and twisting into the sky. "An unattended fire can be a real hazard," said Jack to the growling dog. I feel guilty about spreading this fire, Jack told himself. On the other hand, I haven't forgotten the snake in my bed either.

Flames crackled behind him while Jack returned to his cabin. He was watching from a window when Catrina and Madeline drove onto the driveway of the clubhouse property. An acrid smell of burned wood filled the air as flames continued to flicker across remnants of beams.

"This has gone too far!" exclaimed Catrina angrily to Madeline.

"Jack Kelsey probably set the fire because of the snakes," said Madeline who was despondent about losing the cabin—again. "They were his own snakes."

"If we were as low as Jake and his helpers are, we'd burn their buildings," stated Catrina. "But maybe we won't do that.

We'll rebuild this place. We'll have to keep more people stationed here."

"Are we going to let them get away with what they have done?" asked Madeline.

"Jack Kelsey doesn't have much money," said Catrina. "He might be ruined if his cabin was burned. Jake, however, has a lot o' money. He is also the cause of all this trouble. Jack works for Jake. Max Coker works for Jake. Therefore, our problem is Jake. He doesn't care enough about anything. He just isn't really interested. All this trouble is just for him to make more money so he could retire in comfort. We can help him retire to his comfortable cottage at Sky Lake. We can help him move to the lake by eliminating his house here. I have heard that upstairs, in his house, he has a model of his development project. Getting rid of this model will be a step toward stopping the destruction of the woods."

During the next few days, Jake Sand worked at his model of Sands Estates. While sipping coffee, he checked the miniature buildings and roads. Each division of houses would be a separate, living space, or village, he told himself. He reviewed his full use of all the land. Land seemed to be in unlimited supply. It was vacant land, anyway, ready for developing. We will have self-sufficient communities with comforts for the good life. The people living in such homes will find contentment. I'll get richer—and have abundant resources so I can retire without ever again having to worry about finances. I'll retire to my cottage at Sky Lake.

Catrina has quit, concluded Jake with satisfaction while he continued to check his model. I know the right people. I have the required permits and have the money. Maybe I could have another party to celebrate our good fortune—with the success of Sands Estates. I could even invite Catrina, Madeline and those other losers to show them that I can overlook the trouble they have caused. That would be a fine gesture.

The next morning, keeping up with his reputation, Jake made the usual calls to people inviting them to another party at his place. Feeling excited about the fulfillment of his plans, he drove to the construction company to talk to George Willis. "Need another window, Jake?" shouted George in greeting. "I keep them in stock—lots o' them. You're our best customer—for that particular window."

"We won't be needing any more windows, George," replied Jake. "Thanks for the thought, however. I'm here to invite you, and any of your crew, to another party at my house tonight. This is a celebration. I'm inviting the usual bunch, that my wife calls cronies, and Max Coker with his crew. Of course, Jack Kelsey, is invited. I'm actually going to invite my other neighbors, the naturalists, particularly Catrina and Madeline—to show them I don't hold grudges—there are no hard feelings." Impressing himself more all the time, the more he talked, he added, "It's too bad they lost their clubhouse—again. Maybe our party will make them feel better." He wiped his eyes in mock grief.

"You're just too kind, Jake," laughed George.

"Oh, I know," exclaimed Jake. "Sometimes I just get overwhelmed with how good I am."

"Well," said George, moving glasses and a bottle from behind his desk then going to the cooler, "maybe I could help you recover from feeling so pleased about your yourself."

"That's so good of you," exclaimed Jake receiving a large glass of rum and coke. "Here's to Sands Estates," he shouted, raising his glass to clink against George's uplifted glass. After enjoying a few more drinks, the two men left the building to get ready for the party.

Driving home, Jake stopped at a roadside park by a lake to make more phone calls for his party. He called Madeline Cooper and invited her, along with Catrina and any other naturalist who might like to come to his party—tonight. Talking to Madeline brought Jake's thought back to the forest. I don't know why the environmentalists are so concerned about the woods, he mused. No one goes there anyway. It's just wasted land. It's a mysterious place. Some say it's even haunted. Scares me anyway. People have gone there and not come back. Parts of skeletons have been found. The area should be cleared and made into a decent place for people to live. In that way, I can actually accomplish something. I'm doing what the pioneers did—going out and conquering the wilderness. Catrina would say there is no wilderness left to conquer and we now have to rescue what is left. Well, the naturalists have given up or will do so in a short time. They were wrong anyway. Why would anyone oppose what I'm doing? I've invited the environmentalists to my party to smooth things over. I don't like trouble. Now, I'm just tired. I'll have a nap then I'll be in great shape for my party.

By the time the fire trucks got to Jake's house, the site was a patch of smoldering rubble. While watching from a car as far away as possible, yet still having a view of the site, Catrina said to Madeline, "We arrived early, thinking we might also want to leave early, and no one was at Jake's house. The door was unlocked. The place was deserted. Seeing the model of Sands Estates, we realized that the project was as destructive as any of us had thought. When we crossed those two wires, that unfortunate fire started."

"The model really burned rapidly," exclaimed Madeline. "The fire spread quickly. There was nothing we could do—or, should I say, nothing we wanted to do—except get out o' there."

"I hope Jake didn't lose anything that was important," said Catrina. "I doubt that he would have anything valuable in the house. Jake has lots of money. He can move to his cottage at the lake. He always wanted to go there to retire."

"I wonder what happened to Jake," said Madeline. "Other people arrived later then left. We don't want to be seen around here and we have a conference to attend."

"We should leave," agreed Catrina, starting the car. She drove southward, proceeding to the conference.

When Jake woke up in his car and saw sunlight and heard birds singing, the first thing he thought about was his party. What have I done? he asked himself, incredulously. Have I missed my own party—again? This is ridiculous, he thought, starting his car and proceeding to drive northward. A fragrance from roadside apple blossoms drifted into the car although he could think only about his house.

Driving past the naturalists' property, he was surprised to see a large pile of logs located beside the site of the burned cabin. The clubhouse is going to be rebuilt with logs, he noted. I wonder what they might have done to my house.

His worries turned to a cold grip of fear when he looked for the familiar outline of his house and saw in its place only open space above charred, smoldering ruins. At first he refused to believe the sight; but slowly had to accept it, as he got closer. An acrid smell of burned wood filled the air. Smoke curled away from blackened beams.

He looked briefly for anything that could be salvaged. He found nothing. The model has been destroyed, he told himself. Such destruction will not stop the project itself. The only building left on this property is the old, naturalists' cabin back by the woods. I'm not going to live in a cabin, especially the environmentalists' shack. I'll move to my cottage at Sky Lake. I hope they haven't burned that place. I must go and see if my cottage has survived.

Jake drove north then turned west toward the lake. The road followed an elevated area above a creek. A glimmering section of the lake appeared occasionally in the distance. A larger expanse of water became visible when Jake drove closer to his property. Grayish-blue outlines of islands loomed amid the open expanse of water.

Crows cawed from the top of a white pine when Jake parked his car on the driveway of his home. He was pleased to see the familiar structure. I'm relieved to know they haven't harmed this place, noted Jake, stepping out of his car. I always planned to live here some day anyway.

Entering the building, he was greeted by familiar walls with pine paneling. Interior air was cold and damp. He turned on an oil heater before stepping out onto a screened veranda then going down a stairway to his boathouse. The boat rested quietly on water rippled only by the departure of an otter. Leaving the boat, Jake proceeded to a second stairway bringing him to a patio. From here, he watched the otter swimming next to shore. Farther out, a flock of mallards splashed onto the lake's surface near two loons that seemed to take turns diving for fish.

Jake went to the kitchen where he prepared coffee and carried a cup of the steaming drink to a table beside windows overlooking the lake. He sat down on a chair next to this table, sipped some coffee then telephoned Jack Kelsey, asking him to not undertake any further actions against the naturalists until some point in the future after a time of rest. Jack said that faulty wiring was being reported as the cause of the fire at Jake's house. "Faulty wiring?" exclaimed Jake. "That fire was caused by a faulty wife. However, there isn't anything I can do under the circumstances."

With a second call to Max Coker, Jake requested the projects to be stopped until he got back after a few days of relaxation. A third call went to the police. He asked them to routinely check his cottage because it was so isolated. Having made the calls, Jake sipped more coffee and concluded to himself that he was more comfortable about calling off his projects until he got back to work and could be in charge of things rather than have any loose ends get out of control.

Having made his calls to the outside world, he noticed his fishing pole leaning against a wall. His tackle box was on the floor next to the pole. Maybe I should do some fishing, he thought. He picked up the pole and box then walked down to his boathouse. After untying the boat he pushed it out beside the dock.

With his fishing pole in one hand and tackle box in the other, he started to step onto the boat. His foot had just left the dock and started moving out toward the boat when the ketch on his tackle box released emptying its contents into the water. He lurched to close the open box and caused his foot to miss the boat. He followed his fishing equipment into the water.

Thrashing wildly in panic until he felt his feet touch the bottom, he stood up in waist-deep, chilling water. He retrieved as much equipment as he could find quickly and left it in a pile on the dock. He also pulled the boat back to the dock. Although he was chilled, having wet clothes, he entered his sleek craft and started the motor.

With a whirring of rapidly moving wings, ducks took to flight from the lake's surface near shore. A dark-blue ripple of water curled away from the front of the boat as it left the dock.

Spray flew back from both sides of the sleek craft when Jake drove at full speed, pointing the bow toward grayish-blue islands in the distance. Having sped past the first island wrapped in mist, Jake moved toward open water where there was nothing visible between water and sky. Gulls circled overhead.

When the horizon became indistinct because of strands of fog, he looked back to determine his location. He recognized some landmarks and turned toward them. Stretches of mist drifted in from the lake. They overtook him and seemed to be racing him to the shore. He lost the race, finding himself in a dense cloud of fog.

Jake felt a pulse of raw fear when rocks loomed up out of a bank of mist ahead of the boat. He swerved sharply, almost turning the craft on its side. No crash followed. Jake reduced speed. His heart pounded while his mind groped with the dread of being lost. Like old friends, he finally recognized mist-draped rocks and, from them, he safely, and gratefully, returned to his boathouse.

Although pestered by a few black flies and mosquitoes, he was pleased to feel his feet on the solid dock. With the boat back in the boathouse, he hastened to his cottage and was pleased to once again be inside its familiar surroundings. He had a shower, put on dry clothes then went to the kitchen. He prepared chili, using a mixture of chopped steak, tomatoes, onions, celery and chili powder.

A bubbling mixture of chili con carne filled the kitchen with a spicy aroma that sharpened Jake's hunger. Maybe things are getting somewhat back to normal, he reflected after preparing coffee. A sense of security seemed to be overtaking a previous, empty feeling caused by recent failure, cold and hunger. Maybe the difficult times are over, he observed while pouring coffee into his cup. Next, he filled a bowl with delicious-looking chili.

He was about to sit down and enjoy his feast when he saw something come into view then vanish outside his north window. Taking a closer look, he saw a fox, with its nose down, following a trail leading to the top of on outcropping of rock. The hunter stopped and became silhouetted against a rugged background of white pines.

The chili will stay warm for a while, Jake told himself. I'd like to take a picture of that fox. He removed a camera from its case and rushed outside, closing the screen door behind him.

Jake saw the reddish—orange colored animal move away from the top of the rock. There was no time to take a picture. Like a wisp of fog curling at the rock's base, the fox had drifted away. Jake climbed the rock then walked in the direction the fox seemed to have taken. Hearing muffled growls, Jake saw a bunched up form of a fox struggling amid some bushes. The animal stopped moving and looked at Jake, giving a squirrel time to run for cover in a gully.

Before Jake could position his camera, the fox turned in a flowing blur of reddish-orange color and moved to the top of another rock. Jake snapped a picture of this hunter on the rock above a moving cloud of fog. Trees appeared ghostlike against a white sky. After the hunter had jumped out of view, Jake realized he could see only fog. It was like an intriguing, moving intruder—and surprisingly thick.

I must get back to my cottage, Jake told himself. I have an hot dinner waiting for me. Suddenly, I feel hungry again—and cold. He tried to walk as quickly as possible although he had trouble seeing where he was stepping amid a moving covering of fog. This stuff is fascinating to watch, although I almost can't see well enough to walk, observed Jake. I can see little of the ground near my feet and almost nothing else. Fortunately, I'm close to my house.

After stumbling over a rock, he proceeded more carefully. I did not realize I had come this far, he told himself. Progress kept getting harder rather than easier. He tried to follow stretches of smooth rocks. They were, however, separated by areas of thick brush in a white forest. He gradually became angry because he was taking so long getting back to an hot meal and other comforts of his cottage. His single-minded purpose was to escape from this white trap.

Rain sprinkled through the mist and rattled against trees. I've never liked the forest, raged Jake to himself. Maybe I've always hated it. It's pretty from a distance. Now I'm inside it and I just hate being here. Jake held back a scream as fear—almost terror—stalked him. It was like the fog. I can't panic now, or I'm finished, he warned himself.

Everything in the woods suddenly looked the same to Jake. I can't get lost, he told himself. I'm too close to my cottage. I have to keep moving to find a landmark I recognize then I can get back home. I can't give up. I'm not lost until I stop working at getting out of this trap. I have to find my cottage.

Jake groped onward, single-mindedly fighting to get home—to his familiar place where he was comfortable. His clothes had become soaked as well as ripped. He struggled to the brink of terror. Only the numbing impact of exhaustion forced him to stop. I know, he cautioned himself, that a person, when lost, should stay in one place until rescued. I have refused to accept being lost. Therefore, I fought to get home. Also, the theory of waiting to get rescued is sensible if there was a chance of being rescued. I could be here a long time before anyone found me. Night will soon be upon me. I would hate to have to be here all night. There are more black flies and mosquitoes in the woods than by the lake. I guess wind off the water helps to get rid of the flies. I'm going to keep traveling. If I give up, I won't find my cottage. If I keep looking, I can't get any more lost.

The forest was wet and rocks were slippery. Jake thrashed

onward, fighting the land. He waded through boggy areas along with swamps. He slipped across rocky surfaces. Bruises and scrapes accumulated.

Exhaustion forced Jake to rest at the base of a white pine. I can't accept being lost, he told himself again stubbornly. I also know I am lost. I'm bleeding from scrapes and black fly bites. Mosquitoes are another torment. I'm sweaty, dirty, wet, sore and tired. Fear is tiring. When I built a model of the woods, I had everything in order. I see no order here. I guess this is the first time I've actually been in the forest. Maximum use of space was the guiding idea of my model. Now I'm lost within the confines of my own borders. The real forest is much larger than I thought. My model of the woods started after the area had been cleared. The part of the wilderness I am now in is the part I never wanted to see. It's as bad as I thought it would be. In fact it's worse—much worse. Maybe my poor physical condition has brought on fatigue before I could really hurt myself.

Jake slept soundly at the foot of the tree. He woke up in the grayness of dawn and did not want to face what he saw. The fog has cleared, he told himself. The forest is still out there. I somehow hoped I would find myself back at my cottage. I don't know what to do. I must do something. I can't just sit here. I'm hungry. Maybe the worst parts of this situation are all the discomforts along with not knowing where I am. My clothes are wet and torn. Blood has dried on my skin from black fly bites and scrapes. I'm shaking from the cold. I'm numb with fatigue. I was at first wearied by fear. Now, it seems to give me some energy. I'll keep walking.

Jake wandered throughout the day. At dusk his legs were too unsteady to continue. He sat down to rest at the edge of a vast swamp. He watched a water snake move slowly from grass bordering an expanse of smooth rock sloping to the water. Extending forward like a flow of water, the reptile turned its head toward Jake. A forked tongue flickered from the head before the creature reached the water and slithered away.

I have to get out of here, Jake told himself. I'm too tired to walk any farther. I've had nothing to eat. If I can gather enough energy, I'll use it. At least I don't have anyone worrying about what I'm doing. I don't, therefore have to worry about them. That's at least good news—and maybe the only good news I can think of right now. I just have to concentrate on myself. I have to try again to get out of here.

Jake stood on unsteady legs and stumbled onward. He followed the edge of the swamp then stepped along the top of a beaver dam bringing him to an elevated ridge.

He approached an area of rock topped by moss. Water seeping from the moss flowed over a smooth section or rock. Jake slipped on this wet surface and tumbled down an embankment. When his knee hit a rock, the knee dislocated. It snapped back into place when he rolled in a flood of pain. He fell on his arm, sending a shot of pain through it. A gash on his arm brought blood trickling to his hand before he lost consciousness. He saw himself lying on the forest's floor.

When Jake regained consciousness, he could feel only pain and just see blurred shapes. He sank back into a long sleep. He woke up once and thought he saw a person approaching. Jake found relief from pain as sleep returned.

He woke up beside the warmth of a campfire. There was a bandage on his arm. "Would you like some soup?" asked an old man who sat near the fire.

"Thank you," exclaimed Jake as he received a bowl of soup and a spoon. He slowly savored the delicious, vegetable ingredients. "I've never enjoyed anything so much," he managed to say between spoonfuls. "Thank you for helping me. Who are you?"

"I'm Caleb Pine," answered the man. "I live on Bear Trap Mountain and in the woods around here and I found you. You got lost. You have been lost for a long time."

"You have saved me from this forest," replied Jake.

"I also want to save the forest from you," said the old man whose face had a kind and pleasant glow along with youthfulness that contradicted his gray hair and generally older appearance. The spark of light in his eyes was his most outstanding feature.

"I'm Jake Sand," said Jake, wanting to introduce himself to this youthful, helpful, somewhat old man.

"I know who you are," the man replied. "I know about your development plans for this forest. From your plans, you have left out one thing."

"What have I left out?" asked Jake before finishing the soup.

"You have left out the beauty of the forest," said the man casually.

"There is no beauty in the forest," countered Jake.

"Because you have not seen beauty does not mean that it is not there," said the man. "I'm going to help you."

"You have helped me already," replied Jake, "and I can't thank you enough."

"You have not yet seen the forest because you have not seen the spirit world," explained the old man. "I have awakened you from minor injuries. Now I am going to awaken you again from a deeper injury so you can see where you are. Have a cup of tea."

"Thank you," exclaimed Jake, gratefully accepting the tea.

"Your physical injuries are healing," observed Caleb. "I'm going to help you until you are completely healed. Afterward, I'll show you the way to your house." Pointing to a yellow flower with mottled, green leaves, Caleb said, "Do you see that trout lily?"

"Yes," answered Jake.

"I did not see that lily in your plans for the forest," countered the old man. "Everything is important because each part is connected to every other part. All the pieces of this forest must be preserved in a park for the benefit of all people, not just a few. You were able to build Sands Estates. But could you build one of these lilies?"

Jake just looked at the old man and said, "No."

"This lily," continued Caleb, "is connected to each other part of the forest and the spirit of the forest that is the same spirit that connects to you. Because this spirit is in you too, you know it, you just have not realized it or remembered it for a long time. You must get to know yourself first and then you will see that you also are part of the forest and are an individualized part of the same Creator who is life. When you get to know yourself, you will realize you are connected to the forest through its spirit and its life. You would no longer want to pollute the woods or destroy it any more than you would deliberately want to harm yourself because both are spiritually connected. The lilies are complete in themselves. They don't come here to experience and develop like people do because they are complete just the way they are—like other segments of the forest such as wolves. People are learning and developing—such as you. You have been lost for a long time and are finding yourself again—your spiritual self—the one and only Jake Sand. Each person is unique and has a distinct purpose to work on and try to develop and improve. You, like others, are an unique individual with a purpose set only for you. You have been lost a long time. You must find yourself, and your true path in addition to remembering you are not alone. You are an individualized part of the Creator, with each other part, like one living universe breathing in and out, having nothing new under the sun except the individual's personal experience and development, unique for each one, as each responds differently. That's what it's all about. We are all working together, differently and uniquely. You are connected to the rest of life—including this forest."

Listening to the old man's words, Jake felt them leading him into his surroundings. The surrounding woods brightened as if an intervening veil had lifted showing him he was not separate from the trees or the sky or life—the same life—and he realized that he was not alone or separate—as he had thought previously.

"There are people at Heron Cove," continued Caleb, "who are working to protect this area by forming a park. Among others, there are Tom Speaker, Cal Tomkins and Jedediah Speaker. They are protecting as much land as possible, including my home at Bear Trap Mountain. They also want to invite you to include your land in this forest, where you were going to build your Estates. They will be contacting you. We wanted to talk to you and I have been waiting for the right time. Your cottage was a good place to visit you. My friend helped me find you. She can find anyone in the forest."

"Friend?" asked Jake, sure that the old man was alone.

"Yes," answered Caleb. "She has been here all the time," he added, pointing to the side of Jake.

Turning, Jake was shocked to see a wolf with yellow eyes watching him. "A wolf?" he exclaimed.

"Yes," answered Caleb. "Actually, she's a coyote. Wolves, including coyotes, are the original dogs. You have just met my dog."

"I've met a lot more out here than I ever thought was out here," said Jake, feeling a sense of elation and peacefulness that he knew would last and he would never be the same again.

"I'm going to take you for a walk," said Caleb, "and tell you about the forest. Remember this walk and the trails. When you get back you will be invited to build a new model of the woods and call your work, Bear Trap Mountain. That is the proposed name of the new park. No one is telling you to do anything. I am a guide leading you to you. You have to find—and today have found—your true path or purpose. I know, as I think you now know, the way to proceed in your own way is to be part of this forest and protect it with a park. I am only a guide, like those familiar landmarks along any journey, reminding you that you are on your right, unique path."

"I am going to have a lot of work to do," said Jake as he and Caleb, with his dog, started walking through the forest.

"Yes," replied Caleb, "and enjoying it for the first time in your life."

"At the entrance to the park," said Jake, "we could construct an information center to inform people of what they will be seeing in the forest."

"There are worn, Indian trails that I will show you,"

continued Caleb. "The Indian villages were here first then came European explorers followed by trappers and traders. Settlers tried farming beside a trail that has become the highway beside your burned house. With this land and its people, there is a great story."

Jake followed Caleb along numerous trails. Beside a river, ferns and other vegetation hung down in profusion to the water's edge. Beyond this river, the two travelers came to a swamp. Light dropped, in varying hues, through a canopy of inter-laced branches. An ovenbird called from the overhead canopy. An hermit thrush sang from a concealing tangle of greenery. After drinking water at a stream's edge, a deer looked up and watched the travelers. A ruby-throated hummingbird darted past Jake and Caleb. Songs of birds, particularly robins, blended together into one harmonious voice. Caleb pointed out each aspect of the forest.

A breeze entered the woods and stirred leaves, along with blossoms, moving each of them in varying wedges of light dropping through the tangle of branches. There was a blend of color, scent and sight that opened up, seeming to reveal itself for the first time to Jake. Through the unfolding view, Caleb added descriptions and explanations. "If you take out one ingredient too far," explained Jake, "I can now see that much more is lost. I should be tired; but I'm so interested—for the first time in my life—that I seem to have a new energy—a spiritual energy as you have said—and I am seeing things for the first time. Especially, there is the one that I'm seeing or meeting for the first time and that is myself. I am remembering, and reconnecting with, myself. I can't think of anything more amazing than that. Have I wasted too much time?"

"Never too late to start," said Caleb.

"I can see now that destruction of the environment limits employment," said Jake. "Only an healthy environment creates endless employment possibilities."

"Everybody works," said Caleb. "Everybody wants to work. However, people have to work in an healthy environment in order to be healthy—or even employed—themselves."

At another river, water cascaded down a rock face. Tendrils of tumbling water splashed over rocks before entering the river in a curtain of spray. After a section of rapids, the stream returned to a quieter course wandering between cliffs looking out over a panorama of the forest. Hills cloaked in blue haze lined the horizon. Indentations marked locations of additional rivers flowing from verdant valleys. Evergreens were brightened with the lighter green hues of new growth. Ferns reached gracefully from the earth, carpeting an emerald world. "Many people have to first be defeated, like you were," said Caleb, "before they look up. Each person—like a flower—has a purpose for being here."

Caleb and Jake journeyed to a clearing on an hill facing southwest. Detecting the movement of cars in the west, Jake said, "There must be an highway down there."

"That's the highway in front of your burned house," said Caleb.

"I think I recognize the fields below us," added Jake. "I still have an home on my property. The small cabin I moved to the back near the woods would serve me well."

"Such an house is all you need," agreed Caleb. "A person's home should not mar the environment."

Upon hearing distant hammering that must be coming from the construction of the naturalists' new, log cabin, Jake said, "Sounds carry a long way in the wilderness." Pointing toward the sound of hammering, he said, "Employees at Bear Trap Mountain Park could live in the naturalists' new log building—if they agree to join the project and that seems to be the type of thing they have wanted. The entrance road to the park could start near that log building then proceed to an information center. Visitors could first learn about the contents of the park before entering it."

"The place where we are now standing was the site of a church that later moved to become the stone church on Church Street in Clarksville," said Caleb.

"A log or stone church could be rebuilt on this site," said Jake. "I'll change my development plans and help to build this new park. I'll contact the people you mentioned at Heron Cove. The naturalists will want to help."

"Maybe your wife met you before you did," said Caleb with a smile flashing across his face.

"That's the most amazing thing," exclaimed Jake, "to think that Catrina was right all along. She was actually agreeing with me while I wasn't fighting her but I was actually fighting myself. Maybe my wife saw, or at least got a first glimpse of, the real Jake Sand before I did."

"Welcome home, Jake," said Caleb.

An hawk screamed from a clouded sky. Sunlight broke away from clouds and moved wedges of light across the area. In the moving light, Jake could see the park. Held in illuminating brightness, light drew him to all life, releasing any space between him and the land, leaving him as part of it. "How will I ever go back?" he asked.

"Now that you are home and recognizing the guideposts you aren't going back because you can now be yourself on your individual path," said Caleb.

"I have much to do," said Jake, "and I haven't even started."

"Time doesn't matter," said Caleb. "Only you matter."

Jake and Caleb, with his dog, walked back into the forest. Their journey took them close to Jake's cottage at Sky Lake. While in the last camp, Jake woke up at dawn to find that Caleb, and his dog, had returned to the forest, leaving Jake to proceed to his cottage. On the way back he found his camera. While dusk faded into shadows of nighttime, Jake enjoyed listening to the repeated, clear calls of a whippoorwill.

Within sight of his cottage, Jake camped for the night. He prepared a meal using food given to him by Caleb. He savored some corn bread followed by tea.

Next morning, he watched a light breeze rippling grass and rustling leaves in the forest. Moving grass and leaves shone in sunlight dropped from a mottled sky. Following a long wait, he walked to his cottage.

After entering the building, Jake telephoned Max Coker and told him that development plans had changed completely. No work was to be done until the project had been explained.

Jake left the cottage and drove south to the site of his burned house. He proceeded slowly along the lane to the cabin that had been moved to the edge of the woods. From his car he unloaded food, clothes and a few possessions he had taken from the cottage. He telephoned George Willis and asked him to bring out another work crew to restore the cabin.

George arrived and the crew went to work immediately to repair the building, getting it ready to be a new home. Additional crews were called and they worked together, putting in a well and connecting water and also electrical systems.

After noting that the restoration was taking place rapidly, George said to Jake, "You were away a long time."

"I didn't plan to be away so long," replied Jake. "I spent some time in the forest and got to like it."

"That's unusual," stated George.

"I'm going to live here in this cabin," explained Jake.

"That's unusual," repeated George, thoughtfully. "You once said you would never live in this shack."

"I have changed my mind," said Jake.

"Are you sure you weren't out in the woods too long?" joked George.

"I haven't been there long enough," answered Jake. "I have missed a lot. I've also changed my development plans for the woods." Turning to look at his new home, he said, "If a small cabin is your home, you can be part of the forest."

"None of this is making any sense to me right now," said George. "I'll get back to work," he added before starting to walk back to the building. The repairs continued at a steady pace. By late in the afternoon, Jake had a useable home beside the forest.

After the workers had completed the restoration, Jake said to them, "This is the finest house I've had. Thank you for repairing it—and doing the work so quickly."

"We're pleased that you like the place," replied George before he and the others walked to their vehicles and drove beside the hedgerow, helping to further re-establish a long-abandoned laneway leading to the burned site then to the highway.

Being alone, Jake looked at his new home located at the edge of the forest. The laneway now extended from the highway to the north side of the cabin. South of the building, there was brush-covered land leading to a willow swamp. A field of weeds was in front, facing east. Hedgerows bordered abandoned farmland to the north.

Jake walked west and entered the forest. Noting the number of maples, he thought, next year I could build a sugar shack and make maple syrup along with sugar. There are numerous beech trees here also—and hemlocks.

In a grove of hemlocks, he started a small fire. Enjoying warmth from his fire, he sat down, rested his back against an hemlock trunk and watched the forest. An ovenbird's cheerful song called from some maples. An oriole sang from the bough of a white pine, accompanying a cardinal's whistling call in the distance. This is my house now, reflected Jake. I think George concluded that my stay in the woods had humbled me. He is right.

Jake returned to his cabin and went inside. He checked the structure for the first time. Interior boards had not been painted. There was one main room with a large window next to a door along the east wall. The south wall also contained a window. In the southwest corner, there was a kitchen. Beyond the kitchen, on the west side, there was a washroom next to a window and a back door. A bedroom with bunks filled the northwest corner. A table had been placed beneath a window on the north wall. Jake sat down on a chair beside this table. From here, he could look out through windows and see fields reaching to the swamp in the south. He could also observe abandoned, farm fields to the north, distant hills in the east and a forest at the back. Jake lit an oil lamp located on a shelf above the table. The lamp cast a soft, ochre-colored light across the cabin's interior. Although I have electric lights, I prefer the light and atmosphere provided by an oil lamp, he noted. The wood stove in the center of the main room will heat this place well. I'll build a new model of the woods and I could keep this work on top of the cabinet against the south wall. When I get time, I'll build a stone fireplace beside the window on the east wall. My first task must be to plant the seeds that Caleb gave me.

Using the last of the corn bread and tea supplied by Caleb, Jake enjoyed the first meal in his new home. Afterward, he took a cup of green tea outside and sat on one of the chairs next to the east wall. That weed field in front of me will be the place to plant the seeds, he noted before savoring the tea. I haven't done any planting before. I should be able to remember what Caleb told me. I have a whole, new life ahead of me. I have a second chance. George didn't seem to mind the change.

Jake marked an area of the weed field extending fifty feet east to west and thirty feet north to south. Next he spaded this section, turning over the sandy soil. He worked slowly. He took breaks and watched the crows. They seemed to also live at the edge of the forest.

After the spading had been completed, he used an hoe to break up the largest clumps and remove roots from the field. The soil was then raked. He drove in stakes at the east and west ends before extending line from one stake to another to mark the rows. Following each string, running east to west, an hoe was used to open furrows in the soil. First the corn, beans and squash seeds were planted, followed by seeds for sunflowers.

While checking the freshly planted, sandy, brown earth, Jake thought, my next project will be to build a stone fireplace. I have seen many different fireplaces. I should be able to build one. I have a new interest in such work. I'm sorry I didn't start planting and building a long time ago.

The crows became accustomed to the man working at the cabin. He collected stones and secured them with mortar to build a fireplace on the cabin's east side.

The completed cabin's first visitor was Jack Kelsey. "I thought I should check to see how you were surviving," said Jack as he and Jake watched flames flicker above logs in the new fireplace. "Having you live here by the woods seems unusual. This place suits me. I like your cabin and garden. Your location is excellent. I thought you said you would never live in this building—and it's so close to the woods."

"I have learned to appreciate the forest," said Jake. "I like cabins now also."

Noting the structure and its location, Jack said, "I like this place better than your other houses."

"I called a realtor and I'm selling my property at Sky Lake," noted Jake. "I'm going to invest that money in my new project in the woods. There will be lots of work for you in the new plans. You will be an employee of the new park, Bear Trap Mountain."

"That sounds like work I'd like to do," exclaimed Jack. "Things are not only changing—they're improving."

"I have to call some people at Heron Cove," said Jake. "As of today, though, you are hired for the park."

"What has happened?" asked Jack, his conversation, like his life, being stripped of all formalities.

"I got lost in the woods," answered Jake. "I met Caleb Pine and he helped me."

"I think a park is a good idea," concluded Jack. "I'm going to enjoy that kind o' work. I'll get back to my house now. Thank you for the park work. I'm in such a good mood, I'm going to have to walk off some excitement energy and I haven't had to do that for a long time."

Jack left the cabin and returned to his home while thinking about the changes that had taken place with the development plans and also with Jake Sand.

The next day, Jake hoed his garden while crows frolicked overhead, worrying an owl at the edge of the forest. Satisfied with his work, he prepared coffee then sat on his favorite chair in front of his cabin and sipped coffee contentedly while crows cawed and frolicked at the edge of the forest. This place feels like home, he reflected, and this is the first time I've felt that way about any house I've been in previously. I apparently take a long time to learn things, or see them—or remember them. I keep thinking about Caleb Pine saying my wife understood me before I did. I must call the people at Heron Cove.

Jake entered his cabin. He lit the oil lamp then started a fire in the wood stove. With heat from the stove warming the room, and lamplight brightening the table, he attached sheets of paper then started to sketch a model of Bear Trap Mountain Park. He drew a miniature outline of the naturalists' new log building. He enlarged it to become the park's entrance. Employees can live in this entrance building, Jake told himself. From this structure, he drew a line indicating a road leading north to an extensive information center. People will enter the center before going to the park, he thought as he sketched another structure. Employees can also live here. From such a village, thought Jake as he drew another line representing a road, the road will go up into the hills past the church that will be rebuilt. After proceeding through an elaborate entranceway, visitors will be ready to discover the park. More lines appeared on the paper. Each line depicted a road or trail with campsites. Jack Kelsey is the first person hired for this work, noted Jake. The developer, Max Coker, can be brought in to these new plans. I will be an employee and, from what Caleb explained, the project is being run from a center in Heron Cove. The naturalists don't know about the park. They will be pleased with the project. They always wanted a park. We will all be employees. The land that I own will become part of the project. The naturalists' land and building will also be included. The project must be open during all the seasons.

After an outline of the plans had been completed, he used a few days to enjoy being home—for the first time. He collected firewood and hoed his garden again before an overcast sky brought rain. For some part of each day, he liked to walk in the forest. He found the adjacent fields to also be intriguing. A fire in the wood stove, or fireplace, kept the cabin comfortably warm and dry during wet or cold days. My shack is small, mused Jake. However, because it doesn't obstruct the forest, then the forest becomes part of my house and, as a result, I have a very large and beautifully interesting home. My home is larger than the confines of my cabin.

While sitting on a chair by the table next to the north wall, Jake enjoyed looking through the windows and watching the fields, willow swamp and forest. Flames flickered across logs in the fireplace and were companions to his plans as well as his dreams. I have waited long enough, he declared to himself, one day when the enjoyment of his new situation reached a pinnacle of enthusiasm that could no longer be contained. I hope I haven't waited too long again. Not many people—none that I know of—have had a slower start at life than I have had. At least now, even I know, that time has come to talk to the environmentalists. As they are part of these plans, we will all have to, eventually, talk to each other. That might be a problem because they likely won't want to talk to me.

The following day, Jake stepped out of his cabin and was greeted by a gray dawn. A breeze from the west brought a light rain. Through the grayness, a robin's song rang clearly. Jake enjoyed the robin's song while he walked to the naturalists' property.

People were working on a log building when Jake approached. The first person he recognized was Madeline Cooper. "You'll have trouble burning this place," she stated upon seeing Jake. "There'll be someone living here all the time —for security," she added. "The property is carefully guarded."

"You have a suspicious mind," replied Jake pleasantly.

"I have a good memory too," countered Madeline.

"Since we are neighbors," continued Jake, "I thought I would come over and invite you and the others to visit my cabin. Don't come and burn my house like you did the last time. Just come over for a visit."

"What are you up to now Jake?" asked Madeline.

"I have something interesting to suggest to you," he said.

"I'm sure," she stated. "Thanks for the invitation anyway. We could maybe visit you tomorrow evening—maybe."

"Thank you," replied Jake, pleased that he did not get an immediate rejection. She'll want to check with the others, he thought, as he turned away and started walking back to his cabin. That went as well as I could have hoped. He enjoyed the robin's song. It seemed to fill the morning.

Throughout the rest of the day and much of the following morning, Jake prepared for his guests. He placed chairs next to the walls. He was partially successful at baking blueberry pies. They were to be served along with tea and coffee. Wood was prepared for fires in the stove as well as the fireplace.

The guests arrived suspiciously and were somewhat subdued. They were surprised to be served hot, blueberry pie along with coffee and tea. Flames jumped along logs in the fireplace.

After serving food and drinks, Jake decided to not wait any longer. He walked to the table where he had placed the diagram of the park. He said, "As surprising as this may seem, I was out in the woods the other day. Some will say I was out too long."

"That thought has crossed my mind a number of times," interrupted Catrina amid laughter.

Jake continued to say, "Between myself and the naturalists, we own the land east of here including much of the forest northward as shown by a drawing I've made here at the table. If you agree to the new plan, your log building would become part of an entrance to a park called Bear Trap Mountain. Farther north, there will be an information center that visitors will enter before getting to the trails or campsites. I'm working with a central committee that is already in action at Heron Cove. People from there will be contacting you. The park will own all land and buildings. We will be employees along with Jack Kelsey. Max Coker and George Willis will help with the buildings. If you agree to the overall plan, the rest of the details can be put into place later."

"Why have you changed?" asked Catrina pointedly.

"As I said before," replied Jake, "I went into the forest, I guess for the first time. I met Caleb Pine who told me a few things. Some of the things he told me, I am now telling you. You seem to be agreeing like I did."

"I've tried to tell you a few things too, Jake," exclaimed Catrina, stirring more laughter. "How did you really discover you had been wrong about things generally and the wilderness in particular?"

The guests became quiet before Jake answered, "I've often been told that I didn't know what I was doing especially by my wife." Laughter again continued to put the visitors at ease then Jake said, "Of course, you stop listening after a while, or get hard o' hearing. Since I blamed most things on my wife, I really didn't have much to worry about until I got lost in the woods. Then the sides fell off and the bottom fell out. I was defeated and in a position to listen when I met Caleb Pine. I now understand things I did not know before—or had not remembered previously. The park is a way of sharing this information and, at the same time, keeping the beauty of the wilderness with all its messages."

Jake stopped talking and served more coffee along with blueberry pie. Talk continued until moonlight dropped through a window and spread across the floor. "I guess we are now employees of Bear Trap Mountain Park," said Catrina before all of Jake's guests left the cabin and walked back to the log building.

Jake was tired, although pleased with the success of his meeting. He slept soundly and started working early the next day. He was busy for a few weeks getting his new home well established. Having looked after his home, he walked to the naturalists' building. He met Jedediah Speaker who was put in charge of the correct placement of trails and roads. Tom Speaker, also from Heron Cove, was to help with the information center. He would work with Cal Tomkins who would be responsible for building the church. Although Max

Coker and George Willis would do much of the general construction work, the naturalists insisted that most of the projects, integrally important to the park, would be built by Rose and Slade Sims of the Clarksville Construction Company. Legal work was to be looked after by the Clarksville lawyer, Mat Holden. Finances were to be directed by the bank manager, Harold Kirby. Madeline Cooper was put in charge of personnel and public relations. Jack Kelsey was to look after camping. The managers of the information center were Lily and Callin River. Jake did not want the responsibility of being one of the general managers although he was forced to have this position. Catrina was selected as the other manager.

Jake surprised people with his seemingly ceaseless activity. He eagerly carried out his responsibilities, working with each part of the park.

The naturalists' log building was increased in size to become the entrance structure. Jack Kelsey received a new log home. The information center became a structure of wood, glass and stone that mirrored the surrounding environment. The church was rebuilt with stone.

Jake was generally checking some part of the park's operation. He walked the trails and was pleased to find that they followed the routes pointed out by Caleb Pine. Jake spent an increasing amount of time in the forest.

Jake enjoyed his cabin home. He woke up at dawn when the first robin sang. Having kindled a fire in the stove, he prepared breakfast, sometimes including his favorite—blueberry pancakes. Afterward, he let coffee stir his thoughts and plans for the day. While sipping coffee contentedly, Jake sat in his favorite chair out by the eastern side of his cabin. From this location he could watch the fields along with his garden and the forest with its park. Hawks could occasionally be seen circling above the cabin.

Although each day was special, in its own way, Jake remembered the time when there was a special vibrancy to the morning as if at that time all things harmonized into an indication, tangible only in his awareness, that all things were on course as they should be when each aspect of the day was in line with the best place it could be at that moment. An hawk soared through a mottled sky dropping wedges of sunlight on the land including Jake's cabin and the park. The light changed in hues while moving across the land, highlighting one feature then another. I wish I had understood, or remembered, more things earlier in my life, reflected Jake as he watched the patches of sunlight. I'm always learning something new.

Jake left his cabin and routinely checked his garden while a robin's song rang across the fields and entered the forest. A breeze waved grass in the fields. He walked to the park's entrance building where a log and stone sign announced Bear Trap Mountain Park. The attached building was being completed with the construction of stone fireplaces.

After walking to the information center, he was pleased by its natural appearance as well as the interpreted wilderness inside. From this center, he proceeded to the stone church that had been completed. The site offered a view of much of the area. Into deepening shades of blue, there stretched the forested hills of the woods.

Selecting one of the many trails, Jake started walking into the forest. He came to a stream's banks topped by lilies. He walked to a sandy bank and here he rested while a breeze moved a green world of forest branches beneath an hawk soaring through a clearing sky.

A person approached and Jake soon recognized the features of Lily River. "I was told you had come this way," she said as she sat down on the sandy bank. "You are an hard person to find, Jake. I wanted to tell you that the information center is ready to start receiving visitors."

"I'm very pleased with the center," replied Jake.

"This park is going to be like a light upon an hill," said Lily. "I suppose you know that Madeline and Jack are getting married?"

"Yes," answered Jake. "Catrina and I are going to renew our

vows again—at the same service. A second marriage is appropriate for Catrina and I because I sort o' missed the first one. Catrina knew who I was; but I didn't and it takes two to properly get married."

"In some way," said Lily, "Catrina knew all the time what you know now. I'm pleased that she never quits—never gave up on you although she took a few shots at you to get your attention."

"My attention is difficult to get apparently," said Jake. "I couldn't get it either."

"Welcome home, Jake," said Lily, "and the forest is our home too."

"The park will help people to see that the woods is their home," affirmed Jake.

"I'm always interested in hearing about people, such as you, who have met Caleb Pine," said Lily.

"I have been wanting to talk to you," replied Jake. "Who is Caleb Pine?"

"There is a Mohawk legend," answered lily, "about an old man who came to visit people. This old man looked poor and asked for food or other help. Each time he was rejected until he met a person who offered food as well as lodging. In return for such help, the old man gave information concerning various plants that could be used for healing." Turning to look intently at Jake, Lily said, "In other words, I think Caleb Pine was sent by the Creator or is the Creator. Is there really a difference? Is the difference only in our perception of whom we meet in a day? Caleb Pine walks in this forest—his forest."

# ABOUT THE AUTHOR

Daniel Hance Page is a freelance writer with fifteen, previous books published and others being written. His books are authentic stories filled with action, adventure, history and travel including American Indian traditions and spiritual insights to protect our environment in the smallest park or widest wilderness.

# OTHER WORKS BY DANIEL HANCE PAGE

*LEGEND OF THE UINTAS*

*PELICAN MOON*

*TRAIL OF THE RIVER*

*INDIAN DAWN*

*ARROWMAKER*

*WILDERNESS TRACE*

*THE MAUI TRAVELER*

*TOLD BY THE RAVENS*

*THE JOURNEY OF JEREMIAH HAWKEN*

*MANY WINTERS PAST*

*WHERE WILDERNESS LIVES*

BLACK ROSE
writing™

CPSIA information can be obtained at www.ICGtesting.com
Printed in the USA
LVOW10s0326081114

412585LV00005B/11/P